DESCRIBING
DATA
STATISTICALLY

Charles D. Hopkins

Indiana State University
Terre Haute

CHARLES E. MERRILL PUBLISHING COMPANY
A Bell & Howell Company
Columbus, Ohio

Published by
Charles E. Merrill Publishing Company
A Bell & Howell Company
Columbus, Ohio 43216

Copyright © 1974 by Bell & Howell Company. All rights reserved. No part of this book may be reproduced in any form, electronic or mechanical, including photocopy, recording, or any information storage and retrieval system, without permission in writing from the publisher.

ISBN: 0-675-08820-8

Library of Congress Catalog Card Number: 73-92000

1 2 3 4 5 6 — 78 77 76 75 74

Printed in the United States of America

CONTENTS

Preface		vii
INTRODUCTION		1
1	THE FREQUENCY DISTRIBUTION	5
2	STATISTICS OF CENTRAL TENDENCY	29
3	RELATIONSHIPS WITHIN THE DATA	45
4	STATISTICS OF VARIABILITY	55
5	STANDARD SCORES	65
6	THE NORMAL CURVE	71
7	CORRELATION	83
Appendix 1	Computational Mastery Text	99
Appendix 2	Check Your Understanding	105
Appendix 3	Table 1	114
	Table 2	116
Glossary		117

PREFACE

The preparation of a book causes the author to think deeply about what that one particular book should do and at the same time what it should not attempt to do. Since this book is the outgrowth of a personal need of mine while teaching educational measurement and research courses, it is devoted to that area of statistics usually referred to as "descriptive statistics." The material covered will supply the "stat" needed for introductory courses in psychology, guidance, counseling, educational measurement, and courses in the behavioral sciences for either the undergraduate or graduate level courses.

The primary objective of *Describing Data Statistically* is to provide the experience needed to develop a mathematical understanding of statistical procedures while learning computational skills. This text is designed to free instructors from teaching descriptive statistics during classtime by providing material organized so that students can pursue this aspect of the course on their own. The author hesitates to call his approach "programmed instruction," however some curriculum people would so classify it. I would rather have the instructor and students view this as an investigative trip through some interesting material that has the outcome of developing a meaningful understanding of particular principles that are needed to work with sets of data.

A special effort has been made to play down rote learning of procedures and terms with an eye to the development of conceptual bases for the topics presented and the computational procedures involved in the interpretation of data. *All sets of data hold many messages.* It is the *function of statistics* to present these messages clearly and to assist researchers and measurement people *in picking up the information provided by data.* Going beyond the utilitarian function mentioned above is the study of statistics for the aesthetic reward of pursuing an interesting study. It is hoped that at least some of those who use this book for its utility will also find the aesthetic reward.

CDH

INTRODUCTION

A. This book is organized in a series of frames. You are now reading frame A. Whether you are studying this book as a part of a course assignment or on your own, some of you are concerned about your ability to be successful in this study because you have heard that it is difficult. This probably stems from the reputation of mathematics as being difficult in any form and the feeling that no one really is expected to understand it. Let me reassure you that almost everyone can work here comfortably. The prerequisites are few and simple—average reading ability, ability to add, subtract, multiply, and divide with only a few errors, and *a firm decision to follow the instructions that will guide you through the material.* Some frames will ask you to respond correctly before you continue.

Instructions will be given for a response that agrees with the book and other instructions if you disagree. *Be sure to follow these instructions precisely.* Some frames will not require a response. This frame (A) is an example of a frame without a response. Go to frame B.

B. Sometimes the response will be an answer to a direct question. The answer will require one or more words and will appear directly below the frame.

Examples:
  (a) What continent has no capital cities?
  (b) Where will the answers for frame B appear?

  (a) Antarctica
  (b) Directly below the frame

Go to frame C.

2 Introduction

C. As you can see from the previous frame it is *important to have the answer in the book covered* so that you may think out the response for yourself. Get a piece of heavy paper the size of this page and keep it with the book.

Did you supply yourself with the mask to shield the answers?

Yes. Go to frame D.
No. Go to frame B.

D. Now that you have the mask, keep each page covered below the frame you are reading. Use the heavy horizontal line at the end of the frame as a guide for stopping.

(a) How much of the page should be covered while reading?
(b) What can you use as a guide for the mask?

(a) Everything below the frame being read.
(b) The heavy black line at the end of each frame.

Go to frame E.

E. Some responses will require that you supply a word or words to complete a statement. For these responses *one blank* will appear for *each word* or *set of words* to be supplied. Example:
(a) The capital of Indiana is _____.
(b) The capital of Nevada is _____.

(a) Indianapolis
(b) Carson City

Go to frame F.

F. Some frames will ask that you respond to one of several alternatives provided. Select the response that you think is correct and record the letter that appears before it. Example:
What animal is considered to be the "king of the jungle?"

(a) rabbit
(b) lion
(c) saber-toothed tiger
(d) Humpty Dumpty

(b) lion

Agree: Go to frame H.
Disagree: Go to frame G.

G. The rabbit is a timid animal and would not be able to rule the forest. The saber-toothed tiger is extinct. Humpty Dumpty is a ficticious nursery rhyme character. Many references in writings are made to the lion as being the "King of the Jungle."

The animal known as the "King of the Jungle" is the _____.

Go to frame H.

H. Some responses will be made on paper not supplied in the text. This will include computation, construction of graphs, etc. For these the answers will be provided in the text as usual.

Where will you find the answers to computational problems presented in the frames?

End of frame as usual.

Go to frame I.

I. The organization of this textbook is designed to help you to interact with the material and to help you to develop an understanding of the material and basic concepts involved with the procedures. Your active participation is required. *Be sure to keep the answers covered until you have formulated your response and follow the directions exactly for your response.* You should expect to experience a certain amount of confusion from time to time and this is normal for *you need to wrestle with new concepts to fully understand them.* Have faith in the system and good luck.

Review any of these instructions that you need to and then

Go to frame 1 in the first section titled, "The Frequency Distribution."

1

THE FREQUENCY DISTRIBUTION

1. Rarely, if ever, do data appear in first form with any pattern of organization to help the reader to discern information held by the figures. A basic function of the broad field of statistics is the organization of data into some interpretable form.

Professor Ego obtained the following scores (number of correct items) for an objective test given to his students in Psychology 201.

```
58  49  53  41  53  61  33  51  51  47  42  55  52  45  49
65  48  44  55  51  50  37  38  54  59  52  46  47  45  48
49  53  56  51
```

Notice how difficult it is to pick out high scores, low scores, and relationship of one score to another.

When Connie Coefficient (Professor Ego's graduate assistant) made the above list for recording she asked herself, "Was a score of 52 a good score compared to other scores?"

Without some organization of the scores it is difficult to answer questions of that nature. She decided then that some score organization was needed.

The simplest way to organize the scores is to *order the scores* from highest to lowest (or lowest to highest). This ordering would proceed so that each score value is placed such that all score values less than it would be on one side and all score values greater than it would be on the other side. This procedure is called an *ordering*.

What first step should Connie do to make the above set of scores more meaningful?

---

order the scores

Go to frame 2.

6 The Frequency Distribution

2. After ordering the scores Connie's list looked like this:

   65 61 59 58 56 55 55 54 53 53 53 52 52 51 51
   51 51 50 49 49 49 48 48 47 47 46 45 45 44 42
   41 38 37 33

   Connie could now begin to see how the score of 52 related to the other scores. By counting she found that eleven (or twelve) scores were above 52 and twenty-one (or twenty-two) scores below 52. She could *not* decide how to assign a position to a 52 because that value appeared twice. She finally decided to assign a 12.5 to each 52 since they were the twelfth and thirteenth scores and 12.5 was the middle point for twelve and thirteen.

   Using Connie's procedure what position would be given to a score of 55?

   6.5

   Agree: Go to frame 3.
   Disagree: Review frame 2.

3. What position would be assigned to each of the following scores?

   61   51   45   41

   ---

   2   15.5   27.5   31

   Agree: Go to frame 5.
   Disagree: Go to frame 4.

4. Connie found the middle numbers for each score—since there was only one score of 61 and only one score above 61 she said that 61 must be second and assigned a position number of 2. There were four 51s and their position numbers were fourteen, fifteen, sixteen, and seventeen—the middle is 15.5. The two 45s took positions 27 and 28 with 27.5 being the middle. The score of 41 was the thirty-first score so Connie assigned a 31.

   Go to frame 3.

5. Assign a position number for each of the test scores.

Score	65	61	59	58	56	55	55	54	53	53	53
Position	1	2	3	4	5	6.5	6.5	8	10	10	10
52	52	51	51	51	51	50	49	49	49	48	
12.5	12.5	15.5	15.5	15.5	15.5	18	20	20	20	22.5	
48	47	47	46	45	45	44	42	41	38	37	33
22.5	24.5	24.5	26	27.5	27.5	29	30	31	32	33	34

Agree: Go to frame 7.
Disagree: Go to frame 6.

6. Remember to use the middle of the position numbers for the assigned number. Assign position numbers for this new set of scores.

32 29 27 27 25 23 23 23 22 21

Score	32	29	27	27	25	23	23	23	22	21
Position	1	2	3.5	3.5	5	7	7	7	9	10

Go to frame 5.

7. The position numbers assigned by Connie are known as *ranks* and the process of assignment is called *ranking*. When you assign position numbers to a set of data you have performed a _____. The numbers assigned are _____.

ranking
ranks

Go to frame 8.

8. The above procedure is one of several ways that meaningful position numbers can be assigned and is the one way most acceptable for statistical procedures. When you meet the topic of correlation in chapter 7, you will see why this way is necessary when deriving a correlation using ranks.

When using a rank to report or interpret performance it is important to include the number being ranked as well as the rank itself. A rank of 3 when 4 are being ranked is probably a much different rank from a rank of 3 when 97 are being ranked.

Which of the following ranks of 3 would probably represent higher performance?

(a) rank of 3 out of 4
(b) rank of 3 out of 97

(b) rank of 3 out of 97

Agree: Go to frame 9.
Disagree: Go to frame 5.

9. Another way of organizing data to aid in interpretation and presentation to readers is to build a table which associates the frequency of occurance with each value. To do this you should (1) identify the highest and lowest values

8 The Frequency Distribution

of the distribution, and (2) sequentially list the numbers starting with the lowest score through the highest score.

(a) What lowest and highest values did Connie find for the set of psychology test scores?

(b) What does her list look like?

(a) lowest—33
highest—65

(b) 33 34 35 36 37 38 39 40 41 42 43 44 45
46 47 48 49 50 51 52 53 54 55 56 57 58
59 60 61 62 63 64 65

Go to frame 10.

10. After listing the scores Connie built a table like the one below and associated the frequency for each value (under the column headed f for frequency) with the score value.

Score value	Tally	f
65		1
64		0
63		0
62		0
61		1
60		0
59		1
58		1
57		0
56		1
55		2
54		1
53		3
52		2
51		4
50		1
49		3
48		2
47		2
46		1
45		2
44		1
43		1
42		0
41		1
40		0
39		0
38		1
37		1
36		0
35		0
34		0
33		1
		34

Go to frame 11.

11. Connie did not use the tally column because she had already ordered the scores and had made a ranking of the values. If she had not ordered them and she did not need the ranks, she could have gone from the raw data directly to the table and associated the individual scores to the distribution by tally marks. A tally mark (✓) is used to indicate a frequency in the column.

To go directly from the raw data, identify the highest and lowest scores and set up the table in three columns—Score value, Tally, f. After associating each individual score with the appropriate score (row) in the table with a tally mark, count the number of tally marks for each score and record under the f column. This table is called a *frequency distribution*. Using the above procedure set up a frequency distribution for the following set of scores:

67 69 70 65 73 69 66 71 70 70 67 68

Score value	Tally	f
73	/	1
72		0
71	/	1
70	///	3
69	//	2
68	/	1
67	//	2
66	/	1
65	/	1
		12

Agree: Go to frame 13.
Disagree: Go to frame 12.

12. Guidelines for setting up a frequency distribution:
    (a) locate the highest and lowest scores from the raw data;
    (b) starting with the lowest score, record score values through the highest in a column called Score value;
    (c) associate each obtained score with its appropriate row with a tally mark ( ) under the column Tally;
    (d) count the tally marks for each row and record the total under the column headed f;
    (e) sum the f column and record under the last row called f (frequency).

Set up a frequency distribution table for this set of scores:

89 83 79 84 82 84 85 83
  85 87 88 87 83 84 83 81

10 The Frequency Distribution

Score value	Tally	f*
89	/	1
88	/	1
87	//	2
86		0
85	//	2
84	///	3
83	////	4
82	/	1
81	/	1
80		0
79	/	1
		16

*f values added after tally marks are recorded.

Go to frame 11.

13. Notice that the f column has been summed (added) and 12 recorded as the total number of values for the frequency distribution. The number of values in a frequency distribution is designated by the letter N. The total number of scores (N) for the psychology test is _____.

---

34    $N = 34$

Go to frame 14.

14. When data are organized into a table that lists the number of times each value appears the table is called a _____.

---

frequency distribution

Agree: Go to frame 15.
Disagree: Go to frame 11.

15. With a large range (difference between the highest and lowest scores) this method may be cumbersome and inefficient.

A method that collapses a large range into intervals overcomes these disadvantages.

Use the following guidelines to set up a *grouped frequency distribution*.

    a. Determine the range:

        Highest score (X_H) minus the lowest score (X_L) equals *range*

$$X_H - X_L = \text{range}$$

    b. Divide the range by 10 and by 20:

$$\frac{\text{range}}{10} = \underline{\qquad\qquad} \qquad \frac{\text{range}}{20} = \underline{\qquad\qquad}$$

c. Decide on the interval size by choosing an *odd* (not even) number between the two values determined in step (b). This will be the size of the interval.

d. Set up the table using intervals in place of single values in the first column.

   (1) The lowest interval must include the lowest obtained score, and it must start with a multiple of the interval size.
   (2) Each succeeding interval starts with a multiple of the interval size.
   (3) The highest interval must include the highest score.

This procedure will give you a *grouped frequency distribution* that has a minimum of 10 intervals and a maximum of 20 intervals and a midpoint for each interval that does not involve decimal values. These criteria are generally accepted for grouped frequency distributions.

Using the guidelines above; develop a grouped frequency distribution for Professor Ego's psychology scores. Use the three columns—Interval, Tally, and f.

Interval	Tally	f
63-65	/	1
60-62	/	1
57-59	//	2
54-56	////	4
51-53	//// ////	9
48-50	//// /	6
45-47	////	5
42-44	//	2
39-41	/	1
36-38	//	2
33-35	/	1
		34

Agree: Go to frame 17.
Disagree: Go to frame 16.

16. Steps for setting up the grouped frequency distribution in frame 15:
   a. 65 - 33 = 32
   b. 32 ÷ 10 = 3.2   32 ÷ 20 = 1.6
   c. An odd number between 1.6 and 3.2 is three (3).
   d. The lowest interval must include the lowest score of 33 and start with a multiple of 3; therefore, the lowest interval must be 33 to 35, the next interval 36 to 38, and continue to the interval that includes the highest score of 65.

What interval includes the score of 65?

12 The Frequency Distribution

63 to 65

Go to frame 15.

17. What table should be used with a set of data that has a large range?

---

grouped frequency distribution table

Go to frame 18.

18. What are the minimum and the maximum number of intervals that would be appropriate for most sets of data?

---

10 and 20

Go to frame 19.

19. Another interpretive technique that Connie could use with the test score data utilizes two more columns for the table. One new column is called the *cumulative frequency (cf)* column, and the other column is called the *cumulative percentage frequency (cpf)* column. The table (without the tally column) will now look like this:

Interval	f	cf	cpf
63-65	1	34	100
60-62	1	33	97
57-59	2	32	94
54-56	4	30	88
51-53	9	26	76
48-50	6	17	50
45-47	5	11	32
42-44	2	6	18
39-41	1	4	12
36-38	2	3	9
33-35	1	1	3
	$N = 34$		

Go to frame 20.

20. The procedure for obtaining the values for the *cf* and the *cpf* columns can be used with *any* frequency distribution—*grouped or ungrouped*. The values for the column *cf* are obtained simply by adding the scores in each interval to the value of the *cf* column for the interval below it. Start at the lowest interval and add one (the frequency for interval 33-35) to zero (the frequency for interval 30-32). $0 + 1 = 1$. How were the *cf* values determined for intervals 45-47, 51-53, and 60-62?

---

45 to 47 $6 + 5 = 11$
51 to 53 $17 + 9 = 26$
60 to 62 $32 + 1 = 33$

Go to frame 21.

21. The values in the *cf* column represent the number of scores that fall below the upper limit of that interval. How many scores fall *below the upper limit* of the interval 48-50?   57-59?   63-65?

---

48 to 50	17
57 to 59	32
63 to 65	34

Agree: Go to frame 22.
Disagree: Go to frame 20.

22. The values in the column *cpf* are obtained by dividing each *cf* value by N (the number of scores) and multiplying by 100.

   How were *cpf* values determined for intervals 45-47, 51-53, and 60-62?

45-47	$(11 / 34) \cdot 100 =$ $.32 \cdot 100 = 32$
51-53	$(26 / 34) \cdot 100 =$ $.76 \cdot 100 = 76$
60-62	$(33 / 34) \cdot 100 =$ $.97 \cdot 100 = 97$

Go to frame 23.

23. The values in the *cpf* column represent the percentage of scores that fall below the upper limit of that interval.

   What percentage of scores falls below the upper limit of each of the intervals 48-50, 57-59, 63-65?

---

48 to 50	50
57 to 59	94
63 to 65	100

Agree: Go to frame 24.
Disagree: Go to frame 22.

24. The previous use of the term *upper limit* implies that there must also be a *lower limit.* RIGHT—intervals have *limits, upper* and *lower.* These should be thought of as end-points or the points where the interval terminates. The concept is much like the idea associated with city limits.

14 The Frequency Distribution

Measurement of any kind supplies data that are considered to be continuous. A rock can be weighed (measured) by a set of scales, and a piece of lumber can be measured with a steel tape. Each of these two measurements involves possible error in the way that the expression of measurement in units associates the idea of *nearness*. A rock that is weighed on a set of scales that weighs to the nearest pound must be reported to the nearest pound. A carpenter who measures a board to the nearest foot must use that value. Our carpenter friend may find that we have chosen the wrong unit for him to use. It is not likely that the foot would be appropriate for carpentry. If we would follow the carpenter for a while, we would find him using different units for different tasks. The inch may be a correct choice for the unit when building a fence for a patio, but a much smaller unit would be needed for building cabinets for a kitchen. Whenever the quantification can be reported in one of several units the data are *continuous*. All measurement techniques provide data that are *continuous*. Tests that measure a human characteristic provide data that are *continuous*. Professor Ego measured the characteristic of achievement to the nearest number of correct responses to items that he had on his test. He could have given two points for each right answer, one-half point, or some other value for the items. Professor Ego's test scores represent data that are
_____.

---

continuous

Go to frame 25.

25.   Another type of data is *discrete* data. Only rarely will a student studying in the behavioral sciences use data that are *discrete*. Discrete data are generated by characteristics or variables that are discrete by their nature. Each can be represented only in a particular unit. The number of children in Professor Ego's family can be reported only in positive whole numbers. There is no concept of nearness when counting and reporting the number of children in a family.

The number of wickets in a croquet set represents information that is
_____.

---

discrete

Go to frame 26.

26.   All measurements provide data that are _____.

---

continuous

Agree: Go to frame 27.
Disagree: Go to frame 23.

27. The key to deciding whether your data are *continuous* or *discrete* lies in the basic nature of the variable. Keep in mind that all data are reported in discrete values whether the variables are *continuous* or *discrete*. Even though the underlying variable or characteristic is continuous, all sets of obtained data are recorded as discrete values.

    Classify each of the following variables according to the type of data (continuous or discrete) each would provide.

    (a) number of kittens in a litter
    (b) height of each child in Ms. Parameter's third grade
    (c) pounds of pollutants added to the air by the cars in Terre Haute for one week
    (d) values obtained from the rolling of a die

(a) discrete
(b) continuous
(c) continuous
(d) discrete

Agree: Go to frame 28.
Disagree: Go to frame 24.

28. When weighing rocks the value 61 represents all the values from 60.5 to 61.5 and 62 represents all the values from 61.5 to 62.5. In the same way a score of 61 on Professor Ego's test represents all values of achievement from 60.5 to 61.5 and 62 represents all the values from 61.5 to 62.5. With a more accurate measuring device we could weigh rocks more precisely; with a more sensitive test we could measure achievement in psychology more precisely. In each of the cases cited above the values 60.5 and 61.5 are the *exact lower* and *upper limits* for the interval which is represented by 61, and 61.5 and 62.5 are the *exact limits* for 62.

    What are the exact limits of the interval for the score 63?

62.5 and 63.5

Agree: Go to frame 30.
Disagree: Go to frame 29.

29. Continuous variables are those variables that can assume any value in measurement depending on the reporting procedure. A rock could weigh 18 pounds if the measuring instrument (scales) weighed to the nearest

16 The Frequency Distribution

pound, or the weight of the same rock might be reported as 18.3 pounds if the scale that is used is accurate to the one-tenth (.1) of a pound. The exact limits of the measure of the 18-pound rock would be 17.5 and 18.5. The difference between the upper and lower limits will be equal to one of the measuring units.

What would be the exact limits for a score of 54 on the psychology test?

53.5 and 54.5

Agree: Go to frame 30.
Disagree: Go to frame 28.

30. If the grouped frequency distribution is put on a number line as it appears in frame 15 it would look like this:

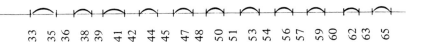

As you can see there are parts of the number line that are not included in our intervals if we use these *apparent limits*.

What could you use to overcome this problem?

---

exact limits

Agree: Go to frame 31.
Disagree: Go to frame 28.

31. In frame 29 we agreed that the exact limits of the interval represented by the score 54 would be one-half of our measuring unit below and one-half above giving exact limits of 53.5 and 54.5.

What would be the exact limits of the intervals 63-65 and 36-38?

---

63 to 65   62.5 to 65.5
36 to 38   35.5 to 38.5

Agree: Go to frame 32.
Disagree: Go to frame 30.

32. Using the exact limits for each interval add a column "Exact" to the table in frame 19.

The Frequency Distribution

Interval Limits		f	cf	cpf
Apparent	Exact			
63-65	62.5-65.5	1	34	100
60-62	59.5-62.5	1	33	97

Interval Limits		f	cf	cpf
Apparent	Exact			
63-65	62.5-65.5	1	34	100
60-62	59.5-62.5	1	33	97
57-59	56.5-59.5	2	32	94
54-56	53.5-56.5	4	30	88
51-53	50.5-53.5	9	26	76
48-50	47.5-50.5	6	17	50
45-47	44.5-47.5	5	11	32
42-44	41.5-44.5	2	6	18
39-41	38.5-41.5	1	4	12
36-38	35.5-38.5	2	3	9
33-35	32.5-35.5	1	1	3
		N = 34		

Go to frame 33.

33. Another way of organizing data is by graphical representation. A graph is usually a two-dimensional picture that utilizes the *vertical* dimension to *represent frequency* and the *horizontal* to *represent some other variable*.

What variable is usually represented by the vertical axis of a two-dimensional graph?

---

frequency

Go to frame 34.

34. Connie built a graph (*histogram*) by using the frequency column (*f*) and the exact limits of the intervals. Her first two intervals were drawn like this:

Complete Connie's histogram, using the data from the psychology test.

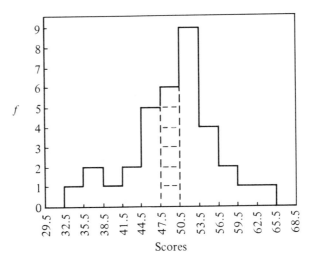

The interval 47.5-50.5 is divided into one block for each frequency to represent the concept of the area reflecting the frequency. The total area under the histogram will equal the total frequency (N).

Go to frame 35.

35. Each graph must have a title that tells what information the graph contains. Example of a graph title:

    NUMBER OF AUTOMOBILES PRODUCED IN THE U.S.
    FROM 1900 TO 1973

    Tables are also named if they are used to present data as part of a study. The title of a graph is usually placed below the graph and the name of a table is usually placed above the table.

    What would be an appropriate title for the graph in frame 34?

    ---

    any title that includes the information presented

    Example:

    SCORES OF 34 STUDENTS ON A PSYCHOLOGY 201 TEST

    Go to frame 36.

36. When building a graph it is important to keep the relationship of the two dimensions so that distortion does not misrepresent the data. It is improper to build graphs out of proportion to reduce large differences or to exaggerate small differences. Generally, a ratio of about 3:5 (three to five) is right for the relationship of the vertical (ordinate) to the horizontal (abscissa).

    (a) What ratio is appropriate for most graphs?

(b) Why might someone use a different ratio when building a graph?

(a) ratio of 3 to 5
(b) to distort the picture to give the viewer a biased view

Go to frame 37.

37. In building the above histogram Connie used the assumption that the scores that fall in each interval are distributed equally throughout the respective intervals. This is necessary because some information is lost in the grouping of scores. The table in frame 32 no longer tells us how many scored particular values. If the six scores that fall in the interval 48-50 were distributed at the points 48, 49, 50, two scores would fall at each point.

How many scores fall at each of the points in the interval 51-53? How many scores fall at 40? at 59?

---

at 51, 52, 53   three (3) each
at 40   one-third
at 59   two-thirds

Agree: Go to frame 38.
Disagree: Repeat frame 37.

38. Another assumption associated with grouped data is that the midpoint in each interval best represents all of the scores in that interval. To find the midpoint of any interval, add one-half of the interval size to the exact lower limit of the interval. For interval 45-47 you would add one-half of three (3/2 = 1.5) to the lower limit of 44.5 (44.5 + 1.5 = 46). What are the midpoints of the intervals in the table in frame 32?

34, 37, 40, 43, 46, 49, 52, 55, 58, 61, 64

Go to frame 39.

39. Using the midpoints as the abscissa (horizontal) values and the frequency for each interval as the ordinate (vertical) values, Connie put a dot above each midpoint for the data from the psychology test scores. The first two intervals look like this:

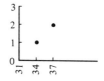

20 The Frequency Distribution

Complete Connie's *frequency polygon*. Notice that the graph starts at the midpoint of the interval below the lowest table interval.

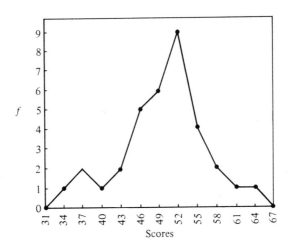

PSYCHOLOGY TEST SCORES FOR 34 STUDENTS

Go to frame 40.

40. If the adjoining dots in the graph are connected by a line the graph is called a *frequency polygon*. A graph that consists of a series of lines connecting points associated with midpoints of intervals is called a _____.

frequency polygon

Go to frame 41.

41. The assumption that the scores in each interval are distributed evenly through that interval generates a graph called a _____.

histogram

Agree: Go to frame 42.
Disagree: Go to frame 37.

42. The assumption that the midpoint of an interval best represents the scores in that interval generates a graph called a _____.

frequency polygon

Agree: Go to frame 43.
Disagree: Go to frame 38.

43. Two other frequency polygons can be obtained by using pairs of columns from the table in frame 32. Using the two sets of values (1) the upper exact limits, and (2) the *cf* column, a *cumulative frequency polygon* can be built. Use the *ordinate for frequencies* and the *abscissa for the upper limits* to build the cumulative frequency polygon for the psychology test scores. (Data from frame 32.)

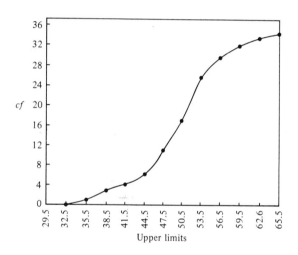

CUMULATIVE FREQUENCY POLYGON
FOR TEST SCORES OF 34 STUDENTS

Go to frame 44.

44. Using the two sets of values (1) upper exact limits, and (2) the *cpf* column, a *cumulative percentage frequency polygon* can be built. Use the ordinate for frequencies and the abscissa for the upper limits to build a cumulative percentage frequency polygon for the psychology test scores. (Data from frame 32.)

22 The Frequency Distribution

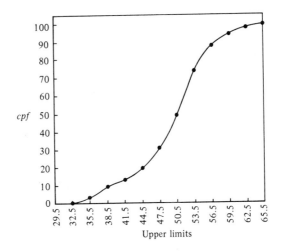

CUMULATIVE PERCENTAGE FREQUENCY POLYGON
FOR 34 STUDENTS

Go to frame 45.

45. Frequency distributions may be presented in two different forms. Within these two forms are several different ways to present the data depending on the information that we use.

    The two forms that a frequency distribution may take are _____ and _____.

---

tabular (table)   graphic (graph)

Go to frame 46.

46. Comparison of a number of frequency distributions reveals that they differ one from another. The two types of cumulative frequency polygons take a characteristic shape like this:

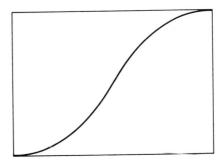

Since the frequencies in the sequence never decrease, the graph will rise as the abscissa value increases. A graph that takes this shape is referred to as an *ogive*.

What is the name given to the particular shape of the cumulative frequency polygons?

---

ogive

Go to frame 47.

47. Other frequency distributions differ in regard to four *properties* or *characteristics*. These four properties—*central location, variability, skewness,* and *kurtosis* are used to describe frequency distributions. When describing a frequency distribution to someone or some audience what properties should you report?

---

central location, variability, skewness, kurtosis

Agree: Go to frame 48.
Disagree: Go to frame 45.

48. *Central location* or central position describes typical performance of the group. For the set of psychology scores Connie might want to know which score was the most popular—the one that occurred more frequently than any other score. Or she might identify a point in the distribution that most typifies all of the scores. In either case some kind of *average* would be used to explain central location. Chapter 2 discusses *central tendency* in more detail.

What would be used to describe a central position of a frequency distribution?

---

an average

Go to frame 49.

49. Variability refers to the amount of dispersion or scattering of scores in the distribution. It is not likely that each score in the distribution will fall at one point—the typical score provided by the average. The more widely the scores in the distribution depart from the average value, the greater the variation or scatter of the scores. Chapter 4 discusses ways of measuring this variation.

What would be used to describe the scattering of scores in a frequency distribution?

---

24   The Frequency Distribution

a measure of variability

Go to frame 50.

50. *Skewness* of a frequency distribution refers to its symmetry or asymmetry. The word "skew" means atypical or not like the rest. If there are some scores in the distribution that are so atypical of the group that the distribution becomes asymmetrical then that distribution is said to be *skewed*. If the atypical scores are high values and to the right (positive direction) in a graph, the distribution is said to be *skewed positively*. If the atypical scores are low values and to the left (negative direction) in a graph, the distribution is said to be *skewed negatively*. If there are no atypical scores and if the two halves of the graph are near mirror images of each other there is *no skewness*.

What skewness is illustrated by each of these frequency distributions?

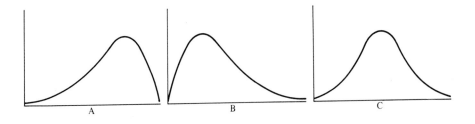

A. negatively skewed (negative skewness)
B. positively skewed (positive skewness)
C. symmetrical (no skewness)

Agree: Go to frame 52.
Disagree: Go to frame 51.

51. Remember that *skew* means *atypical*. The skewness will be in the direction of the atypical scores—those scores that can be thought of as deviate scores.

Go to frame 50.

52. *Kurtosis* refers to the peakedness or flatness of a distribution. A reference for the flatness or peakedness is the familiar bell-shaped distribution that we call the "normal curve." The normal distribution is called *mesokurtic*.

The Frequency Distribution 25

A distribution graph that is flatter than the normal curve is referred to as *platykurtic,* and a distribution that is more peaked than the normal curve is called *leptokurtic.*

Connie learned to report the four properties of distributions and was pleased to add words like "kurtosis," "mesokurtic," "platykurtic," and "leptokurtic" to her vocabulary, but she found it very difficult to drop them in a "social hour" conversation.

Which one of the other properties that have been discussed is closely related to *kurtosis*?

---

variability

Go to frame 53.

53. Supply the missing word(s) for the title of these super-imposed frequency distribution graphs.

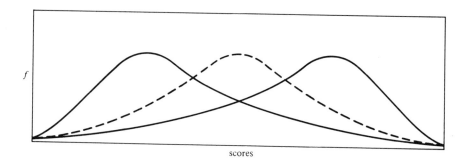

THREE FREQUENCY DISTRIBUTIONS THAT
DIFFER IN _____

---

skewness

Agree: Go to frame 54.
Disagree: Review frame 50.

54. Supply the missing word(s) for the title of the following.

26 The Frequency Distribution

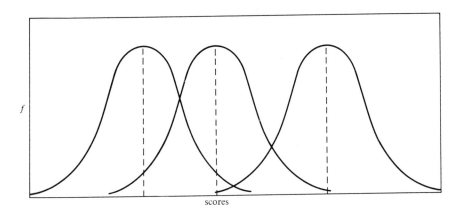

THREE FREQUENCY DISTRIBUTION GRAPHS THAT HAVE THE SAME SHAPE BUT DIFFERENT _____

---

averages

Agree: Go to frame 55.
Disagree: Review frame 48.

55. Two incomplete titles are presented for the following. Supply the missing word(s) to complete the titles.

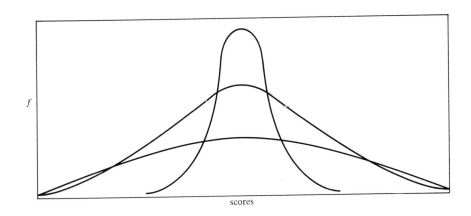

(a) THREE FREQUENCY DISTRIBUTION GRAPHS THAT HAVE THE SAME AVERAGES BUT DIFFERENT _____

The Frequency Distribution

(b) THREE FREQUENCY DISTRIBUTIONS THAT DIFFER IN _____

(a) variability
(b) kurtosis

Go to frame 56.

56. Numerical measures of the four properties (characteristics) are derived through mathematical procedures. Chapter 2 deals with *central tendency* and chapter 4 deals with *variation*. The relationship of two measures of central tendency will be used to indicate *skewness*. A graphical representation of the data and the measure of variation will be used to indicate *kurtosis*. The mathematical measure for kurtosis will be left for a more advanced statistics course.

Go to frame 57.

2

STATISTICS OF CENTRAL TENDENCY

57. When dealing with sets of data a major concern is with the characteristics or properties which are descriptive of the data. The data may be from a group of subjects or may be gathered from one individual subject at different times. In either case, the data could be viewed as a *population* or aggregate. Thirty-four measurements of the length of a tennis court by thirty-four different people would be a population of measurements; thirty-four scores from Professor Ego's Psychology 201 class would be a population of measurements. The immediate concern is deciding what one value most typifies a population.

    Which of the four characteristics that we discussed in chapter 1 would be reported to provide information of this nature?

    central location

    Agree: Go to frame 58.
    Disagree: Review frame 48.

58. An inquiry that might be made about central location is "What value occurred most frequently in the distribution?" When the most frequently occurring score is identified, that value is called the *mode*.

    Using the original set of data from the psychology test, what is its *mode*?

    51

    Agree: Go to frame 60.
    Disagree: Go to frame 59.

30   Statistics of Central Tendency

59. The most efficient way to locate the mode of a frequency distribution is to use a frequency table. Frame 10 has a frequency table for the test scores. Another way to find the mode is to look for the peak of a histogram, or a frequency polygon.

    What score value appears more than any other in frame 10?

    51

    Go to frame 58.

60. The *mode* is the value that is used when the properties of particular subjects fall into categories. Eyecolor is an example of a variable that would fall into categories. The central location is reported through the *mode* since it reports the most frequently occurring classification. To answer the question—"What is the predominate color of eyes for freshman girls at Indiana State University?"—you should use the mode.

    What indicator of central location should be used to report the most popular color for the 1952 Studebaker automobile?

    ---

    mode

    Go to frame 61.

61. Modality is also used to report central location within frequency distributions that have more than one concentration of scores. This might be the case when you have two or more sub-populations in your major population. The distribution of heights in a human population might have two peaks—one the central location of females and one the central location of males.

    A frequency polygon for this population might look like this:

    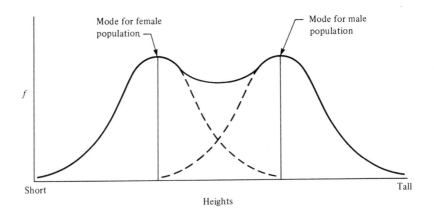

Statistics of Central Tendency   31

A distribution made up of several sub-populations may have several _____.

---

modes

Go to frame 62.

62. For a grouped frequency distribution it is customary to report the *midpoint of the interval* that has the largest frequency as the *mode*.

What is the mode for Professor Ego's test scores after grouping? (See frame 15.)

52

Agree: Go to frame 64.
Disagree: Go to frame 63.

63. Frame 15 has a grouped frequency table that associates frequency of occurrence with intervals. Find the interval that has the largest frequency and locate the midpoint. (See frame 38 for a review of midpoint identification.)

What is the midpoint for interval 51-53?

52

Go to frame 62.

64. A measure of central location associated with those variables that generate data that can be ordered is the *median*. The *median* is that point in a given distribution that has one-half of the cases above it and one-half of the cases below it.

(a) How many cases (N) are in the following distribution?
(b) What is one-half of N?

                4   6   8   10   14

(a) 5
(b) 2½

Go to frame 65.

32 Statistics of Central Tendency

65. Since we need two and one-half cases on each side of the median, let us consider the score of 8.

What are the exact limits for the interval with 8 as a midpoint?

7.5 to 8.5

Go to frame 66.

66. What part of the interval 7.5 to 8.5 would represent one-half of one score?

7.5 to 8.0 and 8.0 to 8.5

Go to frame 67.

67.

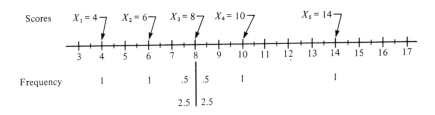

Using the number line above, how many scores fall below 8.0? above 8.0?

What is the name of the point where 50% of the scores fall above and 50% fall below?

2.5   2.5
median

Agree: Go to frame 68.
Disagree: Go to frame 64.

68. What is the median for the scores in frame 64?

8

Agree: Go to frame 69.
Disagree: Go to frame 64.

Statistics of Central Tendency

69. Where the distribution has an even number of scores the median will usually *not* be an occurring value in the distribution. It is customary to divide the middle interval and use the midpoint as the median.

$$4 \frown 6 \frown 8 \frown 9 \frown 10 \frown 14$$
$$\uparrow$$
middle interval

   (a) What is the midpoint of the middle interval?

   (b) How many scores fall above 8.5? below 8.5?

   (c) What is the median?

(a) 8.5
(b) 3; 3
(c) 8.5

Agree: Go to frame 70.
Disagree: Review frame 69.

70. Using the frequency distribution in frame 10 determine the median for the set of psychology test scores.

50.5

Agree: Go to frame 72.
Disagree: Go to frame 71.

71. Counting up the column we find that the score of 50 is in the lower 50% of the distribution and the score of 51 is in the upper 50%.

   What is the upper limit of the interval 49.5 to 50.5?

   What is the lower limit of the interval 50.5 to 51.5?

   What is the median?

50.5
50.5
50.5

Go to frame 70.

34 Statistics of Central Tendency

72. Another way to find the median is to use the *cf* column in a frequency table (see frames 19 and 32). In our set of scores the value $N/2$ appears in the *cf* column and the upper limit can be read directly for the median. The value $N/2$ may not appear in the *cf* column. If it does not it will be necessary to interpolate within the interval to find the median.

    What is the median of this distribution?

    $$4 \quad 6 \quad 8 \quad 8 \quad 8 \quad 10$$

    If the measurements are on the X variable then:

    $$X_1=4 \quad X_2=6 \quad X_3=8 \quad X_4=8 \quad X_5=8 \quad X_6=10$$

7.83

Agree: Go to frame 76.
Disagree: Go to frame 73.

73.

Score	f	cf
10	1	6
9	0	5
8	3	5
7	0	2
6	1	2
5	0	1
4	1	1
	N=6	

This is the way to find the median for the six scores from frame 72.

Using the *cf* column, *identify the interval* where you find the point that has 50% of the scores below and 50% above. Since $N=6$, 50% ($N/2$) of the scores is three (3). The third score falls in the interval 7.5 to 8.5. That interval has three (3) scores, so the median is somewhere in that interval. Using the assumption that the scores are distributed equally throughout the interval, the interval 7.5 to 8.5 looks like this:

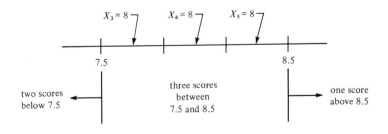

Statistics of Central Tendency 35

The number line looks like this:

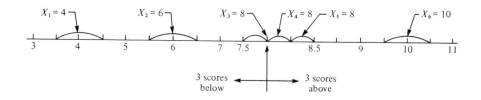

(a) What percentage of the scores fall above the point designated by the arrow?

(b) What percentage of the scores fall below the point designated by the arrow?

(a) 50%
(b) 50%

Agree: Go to frame 74.
Disagree: Review frame 73.

74. (a) What is the numerical value of the point indicated by the arrow?

(b) What is the name of the point indicated by the arrow?

(a) 7.83
(b) median

Agree: Go to frame 76.
Disagree: Go to frame 75.

75. If the point called the median is one-third of the distance from 7.5 to 8.5 then the median must be 1/3 of (8.5 - 7.5) added to 7.5. 1/3 (8.5 - 7.5) = 1/3 (1) = .33̄3 or .33 (nearest hundredth). What is the median value?

---

median = 7.5 + .33 = 7.83

Agree: Go to frame 76.
Disagree: Go to frame 73.

36 Statistics of Central Tendency

76. Find the median for the psychology test scores using the grouped data in frame 32.

50.5

Agree: Go to frame 78.
Disagree: Go to frame 77.

77. $N/2 = 17$. Since we need all the scores in the interval 47.5 to 50.5, the upper limit of that interval is the median. This gives 17 scores above 50.5 and 17 scores below 50.5. Fortunately the median fell at the upper limit of an interval and we could read the value directly from the column "exact limits." If the median had fallen in an interval it would be necessary to interpolate within the interval. (See frames 73 and 74.)

Go to frame 78.

78. Since the frequencies above the median are the same as the frequencies below the median, *the median will divide the frequency polygon area or histogram area into two equal parts.* Refer to the histogram in frame 34 to verify for our data. This also holds for the frequency polygon in frame 39, but is not as obvious as in the histogram.

Go to frame 79.

79. The third indicator of central location is the arithmetical average obtained by dividing the total value of the scores by the number of scores. If we consider the variable being measured as the X variable and Σ to indicate summation then ΣX means to sum the values that we obtained on the X variable. Formally presented, the arithmetical average, or mean ($\bar{X}$), is obtained by formula: $\bar{X} = \dfrac{\Sigma X}{N}$

What name and symbol are used for the arithmetic average of a set of scores?

---

mean $\bar{X}$

go to frame 80.

80. Using X to represent the variable, how is the third score in a distribution symbolized? (See frame 72.)

X_3

Go to frame 81.

Statistics of Central Tendency 37

81. If $\Sigma X = 42$ and $N = 7$, then $\bar{X} =$ _____.

6

Agree: Go to frame 83.
Disagree: Go to frame 82.

82. To find the mean ($\bar{X}$) divide the sum of the scores (ΣX) by the number of scores (N). For these data—to find $\bar{X}$ divide 42 by 7.

$\bar{X} = \Sigma X/N \quad \bar{X} = 42/7 \quad \bar{X} =$ _____.

$\bar{X} = 6$

Go to frame 83.

83. Connie did not find it difficult to obtain the mean of a set of scores. If there was a large number of scores she used an adding machine or a calculator. Connie was more concerned about "What is the mean ($\bar{X}$)?" and "How do the median and mean differ?" She asked Professor Stat if he could help her find answers for her questions. Professor Stat said that the mean of the set of scores was like a balance point where each score provided a force in relation to its distance from the mean. Each score weighed the same but affected the mean in *terms of distance*.

Where would the mean be for these two scores?

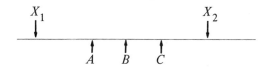

B

Go to frame 84.

84. What is the balance point for these *three* scores?

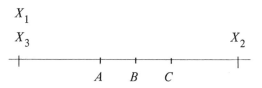

A

Agree: Go to frame 86.
Disagree: Go to frame 85.

38 Statistics of Central Tendency

85. Finding a balance point in a set of data is much like finding positions for the seesaw in a children's playground. The major difference is that in the park the balance point is stationary and we move weights around it. The same concept of weight and distance apply. In finding the mean the distances are within the data and we fit the balance point to the data. If X_1, X_2, and X_3 in frame 84 could be considered as each weighing the same (or pushing equally against the number line) the mean would be placed at the point where the distance from the $\overline{X}$ to X_2 would be twice that from the $\overline{X}$ and X_1 and X_3. The point (A, B, or C) that satisfies this is _____.

A

Go to frame 84.

86. If several (or many) scores are placed on a balance board in the form of a number line, finding the mean is much like shifting a fulcrum along the line until the scores balance around a point. That point is the mean ($\overline{X}$).

What is the balance point for these scores?

X_1 X_2 X_3

.   .   .   .

A   B   C   D   E   F   G   H   I   J   K

D

Go to frame 87.

87. If you were asked to find the average of 142, 144, 146, it is *not* likely that you would add the scores and divide by 3. Connie said that she could just look and see that the mean is 144. Professor Stat said that she used the same principle here that was used in frame 86, and that Connie had used a basic property of the mean to find 144. Connie agreed after thinking about it—"because the 142 and 146 sort of cancel out."

A more formal statement of this property is: "The sum of the deviations of a set of scores from the mean is equal to zero (0). $\Sigma(X - \overline{X}) = 0$.

$$142 + 144 + 146 = 432 \ (\overline{X} = \Sigma X/N = 432 \ / \ 3 = 144)$$

$$146 - 144 = 2$$
$$144 - 144 = 0$$
$$142 - 144 = -2$$
$$\Sigma(X - \overline{X}) = 0$$

Statistics of Central Tendency

What is the sum of the deviation scores, $\Sigma(X - \bar{X})$, for the following distribution?

            9   10   12   13   16

$\Sigma(X - \bar{X}) = 0$ 9 + 10 + 12 + 13 + 16 = 60
                          60 / 5 = 12 = $\bar{X}$

16 - 12 =   4
13 - 12 =   1
12 - 12 =   0
10 - 12 = -2
 9 - 12 = -3
           ——
           0 = $\Sigma(X - \bar{X})$

Go to frame 88.

88. For hand computation summing the scores from raw data may be quite time consuming if you have a large number of scores. If you have a large number of cases (large N) use a frequency table to calculate the mean. By multiplying each raw score value by the frequency of occurrence and summing you can get the ΣX value quickly and probably more accurately.

Using the information from frame 10 you will find:

X	f	$f(X)$
65	1	65
61	1	61
59	1	59
58	1	58
56	1	56
55	2	110
54	1	54
53	3	159
52	2	104
51	4	204
50	1	50
49	3	147
48	2	96
47	2	94
46	1	46
45	2	90
44	1	44
43	1	43
41	1	41
38	1	38
37	1	37
33	1	33
	$N = 34$	$1689 = \Sigma fX = \Sigma X$

$\bar{X}$ = 1689 / 34
$\bar{X}$ =   49.68

Go to frame 89.

40 Statistics of Central Tendency

89. If the data have been grouped as in frame 19, you can use the midpoint as the computational value for each interval. Weight each midpoint value by the frequency of the interval by multiplying. Then sum through the intervals for the ΣX value. Optional: determine the mean for the grouped data of psychology test scores (frame 19).

Class interval	Midpoint X	Frequency f	fX
63-65	64	1	64
60-62	61	1	

$\Sigma X = 1687$
$\overline{X} = 49.62$

Class interval	Midpoint X	Frequency f	fX
63-65	64	1	64
60-62	61	1	61
57-59	58	2	116
54-56	55	4	220
51-53	52	9	468
48-50	49	6	294
45-47	46	5	230
42-44	43	2	86
39-41	40	1	40
36-38	37	2	74
33-35	34	1	34
		N = 34	$\Sigma X = 1687$

$\overline{X} = \Sigma X/N = 1687 / 34 = 49.62$

Go to frame 90.

90. The true mean of the distribution is 49.68 (see frame 88). The difference between 49.68 and the mean estimated from grouped data represents a "grouping error." In general, the grouping error will tend to be less for large numbers of frequencies and greater for small numbers of frequencies.

The grouping error for the true mean and the estimated mean for our data is _____.

49.68 − 49.62 = .06

Go to frame 91.

91. Determine the *mean and median* for each of the two distributions below.

Distribution A	10	12	15	16	22
Distribution B	10	12	15	31	32

Statistics of Central Tendency 41

	A	B
Mean	15	20
Median	15	15

Go to frame 92.

92. What numerical relationship do the two medians of distributions A and B have?

___

same value (15)

Go to frame 93.

93. What numerical relationship do the two means of distributions A and B have?

___

mean of B is greater than the mean of A

Go to frame 94.

94. How can you explain why the *medians* of A and B are the *same* but the *means* are *different*?

___

The mean involves the variable of distance but the median neglects distance and is determined by merely identifying whether a score is below or above a particular point.

OR

Calculation of the mean involves all of the *values* of the distribution scores but determination of the median requires only placing each score above or below a particular point.

Go to frame 95.

95. What positional relationship would the mean and median have in a perfectly symmetrical frequency distribution?

___

They would both fall at the same point.

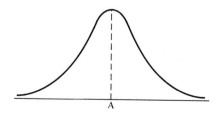

42 Statistics of Central Tendency

Point *A* is that point in the distribution that divides the area into two equal parts, therefore *A* is the median.

Point *A* is that point where the frequency distribution would balance, therefore *A* is also the mean.

Go to frame 96.

96.

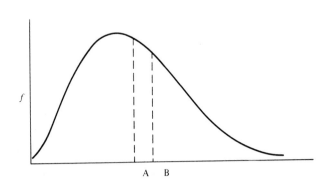

(a) Which average (mean or median) is more affected by extreme scores?

(b) What positional relationship would the mean and median have in a distribution that is skewed positively?

(c) The mean is ___(A or B?)___ and the median is ___(A or B?)___ .

(a) mean
(b) mean is in the positive direction (to the right); median is to the left
(c) $\bar{X}$ is B; median is A

Go to frame 97.

97.

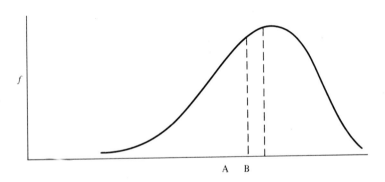

Statistics of Central Tendency

The mean is _____, and the median is _____.

---

mean is *A*; median is *B*

Go to frame 98.

98. If the median and mean of a distribution have different values the skewness will be in the direction of the mean. If the mean is to the right of median the distribution is skewed _____.

   If the mean is to the left of the median, the distribution is skewed _____.

---

positively
negatively

Go to frame 99.

99. The set of psychology test scores has a mean of 49.68 and a median of 50.50. That distribution is skewed _____.

---

negatively

Agree: Go to frame 100.
Disagree: Go to frame 91.

100. The most frequently occurring score in a distribution is called the _____.

   The arithmetic average of the occurring values in a distribution is called the _____.

   The point that is midway through a set of scores is called the _____.

---

mode
mean
median

Go to frame 101.

Statistics of Central Tendency

101. The mode, median, and mean are all statistics that describe a frequency distribution's _____.

---

central location

Go to frame 102.

3

RELATIONSHIPS WITHIN THE DATA

102. Interpreting a set of data involves more than locating measures of central tendency in the data. Connie's original question—"Was the score of 52 a good score compared to other scores?"—has been only partially answered. (52 is greater than the mean of 49.68.) We also know that those who scored 52 scored higher than 50% of the students because 52 is greater than the median (50.5). Any number of questions can be asked in regard to percentages of scores.

Did the student who scored a 52 on the psychology test surpass 35% of the other students? 45%? 55%? 65%? 75%? 85%? 95%?

We are *not* in a position to answer all of these questions, but two answers seem rather obvious.

35%—yes. If 52 surpasses at least 50%, then it must surpass 35%.
45%—yes. Same reason.

55%, 65%, 75%, 85%, and 95% will be left unanswered for now. (We will answer these later.)

Go to frame 103.

103. Let us check for the 75% question first. If 50% is one-half of the way through the distribution, then 75% is _____ of the way through, and 25% is _____ of the way through.

46 Relationships Within the Data

3/4
1/4

Go to frame 104.

104. If we place these three points on a number line they will divide the distribution into _____ parts.

4

Agree: Go to frame 106.
Disagree: Go to frame 105.

105. If the median divides a distribution into two parts, how many parts will three points give?

4

Go to frame 104.

106. If we start at the lowest value and continue to the highest value of the distribution and place the interval for the distribution on a number line it will look like this:

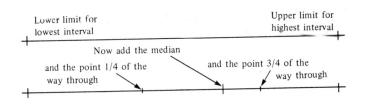

These three points have divided the distribution into four sections and each contains _____ percent of the scores.

---

twenty-five

Go to frame 107.

107. The name given to each one of the three points that divide a distribution into four parts is *quartile*.

A quartile is one of three points in a distribution that divide that distribution into four equal parts.

If Q_1 is the point below which 25% of the cases lie and the median (Q_2) is the point below which 50% lie, then Q_3 is the point below which _____ percent lie.

---

seventy-five

Go to frame 108.

108. A decile is one of nine points which divide a distribution into ten equal parts.

Another name for the fifth decile is _____ .

---

median or Q_2

Agree: Go to frame 110.
Disagree: Go to frame 109.

109. If the first decile (D_1) is that point below which 10% of the cases lie and the second decile (D_2 is that point below which 20% of the cases lie, then the fifth decile (D_5) is the point below which 50% of the cases lie. What other point also divides the distribution so that 50% of the cases lie below?

---

median, Q_2

Go to frame 108.

110. A percentile is one of 99 points in the distribution that divide the distribution into 100 equal parts.

What is another name for the fiftieth percentile (P_{50})?

---

median, second quartile (Q_2), fifth decile (D_5)

Go to frame 111.

48 Relationships Within the Data

111. (a) What is another name for Q_3? Q_1?
 (b) What is another name for D_4? D_6?
 (c) What is another name for P_{20}? P_{80}? P_{25}?

(a) P_{75} P_{25}
(b) P_{40} P_{60}
(c) P_2 D_8 Q_1

Agree: Go to frame 112.
Disagree: Go to frame 104.

112. To find these score values in the distribution you would follow the procedure for finding the median except the proportion would be different. Let us list the steps for finding any particular point in the distribution.

   1. Using a frequency distribution (grouped or ungrouped) compute the cumulative frequencies.

   2. Determine the number of cases that fall below that desired point (pN; p is the proportion shown by the percentile).
   If $p = .50$ and $N = 34$, then $pN = 17$ ($.50 \cdot 34 = 17$).

   3. Find the interval where the desired case falls. Use the cf column and determine the exact limits.

   4. Interpolate within the interval to find the scale value. This is the point that you are looking for.

Go to frame 113.

113. The following formula condenses these steps for efficient computation:

$$P_p = L + \left(\frac{pN - f_b}{f_w}\right) i, \text{ where}$$

L is the exact lower limit of the interval containing P_p
N is the total number of cases
i is the interval size
p is the proportion of cases below the point
f_b is the number of cases falling below L
f_w is the frequency of the interval containing P_p

What is the special formula for finding the median?

Relationships Within the Data 49

$$P_{50} = \text{median} = L + \left(\frac{.50N - f_b}{f_w}\right)i$$

Agree: Go to frame 114.
Disagree: Review carefully frame 113.

114. What is the special formula for finding Q_1?

$$Q_1 = L + \left(\frac{.25N - f_b}{f_w}\right)i$$

Go to frame 115.

115. Using the set of psychology test scores, find Q_1 and Q_3.

$$Q_1 = P_{25} = 44.5 + \left(\frac{8.5 - 6}{5}\right)3 = 44.5 + \left(\frac{2.5}{5}\right)\left(\frac{3}{1}\right) = 44.5 + \frac{7.5}{5} = 46$$

1. Find the proportion of N shown by P_{25}
$$pN = (.25)(34) = 8.5$$

2. Use the *cf* column to find which interval contains P_{25}.
Since 6 scores lie below 44.5, P_{25} is above 44.5.
Since 11 scores lie below 47.5, P_{25} is below 47.5.
Since 8.5 is between 6 and 11 then 8.5 is between 44.5 and 47.5, the limits of one of our intervals.

We can now get our L and f_b values.
L is the lower limit of interval 44.5 - 47.5
f_b is the *cf* value for interval below identified interval (6)

3. f_w is frequency of identified interval (5)
4. Interval size is three (3).

$$Q_3 = 50.5 + \left(\frac{25.5 - 17}{9}\right)3 = 50.5 + 2.83 = 53.33$$

1. $.75 \cdot 34 = 25.5$
2. L is the lower limit of interval 50.5 - 53.5.
   f_b 17 (frequency of intervals below)
3. f_w is 9 (frequency within the interval)
4. Interval size is 3.

50 Relationships Within the Data

$Q_1 = 46.00$
$Q_3 = 53.33$

Agree: Go to frame 116.
Disagree: Go to frame 113.

116. Find D_4 and P_{92}.

D_4 48.80
$P_{92} = 58.42$

Agree: Go to frame 117.
Disagree: Go to frame 113.

117. In the preceding frames we calculated score values associated with particular positions. We could also calculate the positions associated with particular _____.

---

score values

Go to frame 118.

118. In fact, you have already calculated the positions associated with the values of the psychology test scores for the upper limits of each interval. The *cpf* column places the distribution into a scale of 100 units. The *percentile rank* is the value in the 100 unit scale associated with particular score values. What *cpf* value is associated with a score value of 41.5? 56.5? (See frames 19 and 32.)

12
88

Agree: Go to frame 120.
Disagree: Go to frame 119.

119. The frequency table in frame 32 has a *cpf* column that gives the position of the upper limits in the scale 0 to 100.

Read the *cpf* column for 41.5 and 56.5.

12
88

Go to frame 118.

Relationships Within the Data

120. Scores do not fall at the upper limits of intervals, therefore we must interpolate within the intervals to get *percentile ranks* for obtained scores. What is the *percentile rank* for the score of 52? 37? 65?

$PR_{52} = 63$ $PR_{37} = 6$ $PR_{65} = 99.5$ or 100
63% of the   6% of the   100% of the scores fall
scores fall   scores fall   at or below 65
at or below   at or below
52   37

Agree: Go to frame 122.
Disagree: Go to frame 121.

121. Use just the upper limit column and the *cpf* column from the table in frame 32. Let's look for PR_{52}:

> *cpf* value for 53.5 is 76
> *cpf* value for 52 is ??
> *cpf* value for 50.5 is 50

52 is the midpoint between 50.5 and 53.5. The *cpf* value for 52 is the midpoint for the interval with limits of 50 and 76.

$$\frac{76 - 50}{2} + 50 = \text{midpoint of 50 and 76}$$

$$\frac{26}{2} + 50 =$$

$$13 + 50 =$$

$$63 = cpf \text{ value } (percentile\ rank)$$

Go to frame 120.

122. Using the ordinal nature of data we have found ways of viewing relationships within the data. Another way of gaining insight to relationships in data is in regard to the distance of points from the mean. Since few scores fall exactly at the mean of a distribution it is possible to get a value that tells the distance of a score from the mean *and* the direction. If the mean is subtracted from a score that value is called a *deviation score*. The digits of a deviation score give the distance of a value from the mean—the sign of the deviation score gives the direction.

What is the deviation score for a score of 52 on the psychology test?

Deviation score = $(X - \bar{X}) = 52.00 - 49.68 = {}^+2.32$

Go to frame 123.

52 Relationships Within the Data

123. A deviation score of $^+2.32$ tells us that the raw score is 2.32 units above the mean.

    (a) How do you interpret a deviation score of $^-2.68$?

    (b) What is the value of the raw score associated with a deviation score of $^-2.68$?

(a) the score is 2.68 units below the mean
(b) 47

Go to frame 124.

124. A deviation score can be found for each score in a distribution. What is the sum of any set of deviation scores?

$\Sigma(X - \bar{X}) = 0$ (See frame 87.)

Go to frame 125.

125. What is the value of a deviation score for a score that falls at the mean?

---

0   $(\bar{X} - \bar{X}) = 0$

Go to frame 126.

126. (a) Find $\bar{X}$ for the following distribution.
    (b) Find a deviation score for each obtained value.

    60 57 64 58 57 59 62 61 62

(a) 60
(b)

Scores	57	57	58	59	60	61	62	62	64
Deviation scores	$^-3$	$^-3$	$^-2$	$^-1$	0	$^+1$	$^+2$	$^+2$	$^+4$

Go to frame 127.

127. (a) What is the arithmetical average for the set of deviation scores? (frame 126)

    (b) On the average, how far are the obtained scores from the mean?

(a) 0 $\Sigma(X - \bar{X}) = 0$ $\dfrac{\Sigma(X - \bar{X})}{N} = \dfrac{0}{9} = 0$

(b) 2   If all nine scores had been 60, then "on the average" they would be 0 units from the mean. Obviously our scores do vary. Two (57 and 57) scores are 3 units from the mean, 3 scores (58, 62, 62) are 2 units from the mean, 2 scores (59 and 61) are 1 unit from the mean, and 1 score (64) is 4 units from the mean. The total for deviations is 18. To find an average we divide by N.

$$18/N = 18/9 = 2$$

Go to frame 128.

128. When you are concerned about absolute distances and disregard the direction of the deviation scores you obtain the *average deviation,* or mean deviation. This procedure treats the two values 59 and 61 the same because they are, indeed, each one unit from the mean of 60. Likewise, 58 and 62 are treated the same since each is two units from the mean.

To find the average deviation of a set of scores find the deviation scores for the set, treat them all as positive values, sum, and divide by N.

Formally stated by formula:

$$\text{average deviation } AD \text{ or mean deviation } MD = \dfrac{\Sigma |X - \bar{X}|}{N}$$

The two perpendicular lines that replace the parentheses indicate that you should disregard the sign of each deviation score. Use the *absolute* value.

Find the average deviation (AD) for this set of scores:

6   8   9   9   11   12   12   13   15   15

54 Relationships Within the Data

$AD = 2.4$

$\Sigma X = 110 \quad N = 10 \quad \overline{X} = 110/10 = 11.00$

$$
\begin{array}{r}
|15 - 11| \cdot 2 = 8 \\
|13 - 11| \cdot 1 = 2 \\
|12 - 11| \cdot 2 = 2 \\
|11 - 11| \cdot 1 = 0 \\
|9 - 11| \cdot 2 = 4 \\
|8 - 11| \cdot 1 = 3 \\
|6 - 11| \cdot 1 = 5 \\
\hline
N = 10 \quad 24 = \Sigma |X - \overline{X}|
\end{array}
$$

$$AD = \frac{\Sigma|X - \overline{X}|}{N} = \frac{24}{10} = 2.4$$

Go to frame 129.

---

129. The average deviation is used rarely since there is seldom need for this statistic. This discussion was included primarily as a step to the next section on measures of variability.

Go to frame 130.

130. Can you now answer the questions that we left unanswered in frame 102?

    Did the student that scored 52 on the psychology test surpass 55% of the other students? 65%? 75%?, 85%?. 95%?

55%   yes.
65%, 75%, 85%, 95%   no.

$PR_{52} = 63$ 63% of the scores fall at or below 52

Go to frame 131.

4

STATISTICS OF VARIABILITY

131. Another dimension to the question asked by Connie—"Was the score of 52 a good score compared to other scores?"—involves the scattering or dispersion of scores in a distribution. To describe a frequency distribution, or to interpret scores, the property of variability must be given attention.

    What two statistics of variability have been presented in previous chapters?

    ---

    range (frame 15), average deviation (frame 128)

    Go to frame 132.

132. The range is *not* used much as a measure of variability since only two scores contribute to the range. For review— the *range* is the difference between the high score and low score for the distribution.

    What is the range for the distribution of psychology test scores?

    ---

    range = $X_H - X_L$ = 65 - 33 = 32

    Go to frame 133.

133. What is the average deviation for the following set of values?
     55 57 59 62 63 64

    AD = 3.00

    Agree: Go to frame 134.
    Disagree: Review frames 126, 127, 128.

56 Statistics of Variability

134. One reason that the average deviation is little used is that the use of absolute values (disregarding deviation score signs) is to be avoided in statistical work, if possible. The use of absolute values resulted from the fact that the deviation scores always sum to zero. Another way to overcome this problem is to use a squaring of values. Whenever a value [whether positive (+) or negative (−)] is squared the resulting value will be positive.

What is the sum of the squared values of the deviation scores for the set of scores in frame 133?

---

64    $\Sigma X = 360$ $N = 6$ $\bar{X} = 360/6 = 60$

X	$\bar{X}$	$(X - \bar{X})$	$(X - \bar{X})^2$
64	60	4	16
63	60	3	9
62	60	2	4
59	60	−1	1
57	60	−3	9
55	60	−5	25
		0	$64 = \Sigma (X - \bar{X})^2$

Go to frame 135.

135. What is the average of $(X - \bar{X})^2$ values?

---

average of $(X - \bar{X})^2$ values $= \dfrac{\Sigma(X - \bar{X})^2}{N} = \dfrac{64}{6} = 10.67$

Go to frame 136.

136. The value 10.67 is the average size of the squares that we obtained by squaring the deviation scores. The 10.67 is a *mean square*. Notice that the mean square is a two-dimensional statistic. The value 10.67 is the *variance* of this set of scores.

(a) What statistic is defined as the sum of squares of deviation scores $[\Sigma(X - \bar{X})^2]$ divided by the number of scores (N)?

(b) What is its formula?

(a) variance (σ^2)

(b) $\sigma^2 = \dfrac{\Sigma(X - \bar{X})^2}{N}$ If $(X - \bar{X}) = x$, then $\sigma^2 = \dfrac{\Sigma x^2}{N}$

(X—symbol for a raw score; x—symbol for a deviation score)

Go to frame 137.

137. A table for the computation looks like this:

Score X	Mean $\bar{X}$	$(X - \bar{X})$	$(X - \bar{X})^2$
64	60	+4	16
63	60	+3	9
62	60	+2	4
59	60	−1	1
57	60	−3	9
55	60	−5	25
		0	64

Graphically it looks like this:

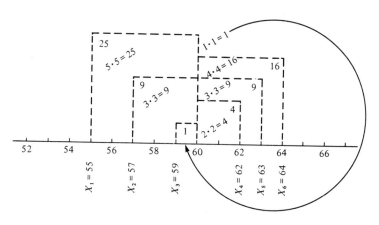

The six squares have an area equal to 64.

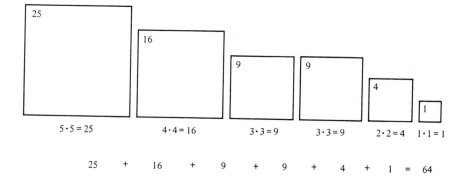

$25 + 16 + 9 + 9 + 4 + 1 = 64$

58 Statistics of Variability

The average size square (the *mean square*) will have an area of 64/6 or 10.67.

[10.67]

(a) What value multiplied by itself equals 10.67?
(b) What is the length of one side of the mean square? $\square \cdot \square = 10.67$

(a) $\sqrt{10.67}$ or 3.27
(b) 3.27

Go to frame 138.

138. What is the square root of σ^2? ($\sqrt{\sigma^2}$ = _____)

$\sqrt{\sigma^2} = \sigma$

Go to frame 139.

139. The square root of the variance is called the *standard deviation* (σ).

Another measure of variability that is the square root of the variance is called the _____ .

---

standard deviation

Go to frame 140.

140. Calculate the standard deviation (σ) for this set of values:

9   11   11   13   16

[The standard deviation (σ) for five values is not really useful but is used here for pedagogical reasons.]

$\sigma = 2.37$ $\Sigma X = 60$ $N = 5$ $\bar{X} = 12$

X	$\bar{X}$	$(X - \bar{X})$	$(X - \bar{X})^2$
16	12	4	16
13	12	1	1
11	12	-1	1
11	12	-1	1
9	12	-3	9
		0	28

$$\sigma = \sqrt{\frac{\Sigma (X - \bar{X})^2}{N}} = \sqrt{\frac{28}{5}} = \sqrt{5.60} = 2.37$$

Go to frame 141.

141. The defining formulas in frames 136 and 139 are *not* efficient formulas for larger numbers of cases. An identity, or equivalent formula, used for computation of the variance is:

$$\sigma^2 = \frac{\Sigma X^2 - \frac{(\Sigma X)^2}{N}}{N}$$

and, the standard deviation is

$$\sigma = \sqrt{\frac{\Sigma X^2 - \frac{(\Sigma X)^2}{N}}{N}}$$

Compute the variance and standard deviation for the set of data from frame 140 using the above computational formulas.

X	X^2
16	256
13	169
11	121
11	121
9	81
$\Sigma X = 60$	$\Sigma X^2 = 748$

$$\sigma^2 = \frac{748 - \frac{(60)^2}{5}}{5}$$

$$\sigma^2 = \frac{748 - 720}{5}$$

$$\sigma^2 = \frac{28}{5}$$

$$\sigma^2 = 5.60 = \text{variance}$$

$$\sigma = \sqrt{5.60} = 2.37 = \text{standard deviation}$$

Agree: Go to frame 143.
Disagree: Go to frame 142.

60 Statistics of Variability

142. If you did *not* get the book answer your mistake might be because of confusion of two similar (but different) parts of the formula.

$$(\Sigma X)^2 \neq \Sigma X^2$$

ΣX^2 = sum of squared scores (square first)
$(\Sigma X)^2$ = sum of scores squared (sum first)

See computation for frame 141.

Go to frame 141.

143. Determine the variance (σ^2) and standard deviation (σ) for the set of psychology scores. (Optional frame—the calculation appears in frame 144.)

$\sigma^2 = 43.87 \qquad \sigma = 6.62$

Agree: Go to frame 145.
Disagree: Go to frame 144.

144. Most computation in statistics can best be organized in a table.

X	f	fX	X^2	fX^2
65	1	65	4225	4225
61	1	61	3721	3721
59	1	59	3481	3481
58	1	58	3364	3364
56	1	56	3136	3136
55	2	110	3025	6050
54	1	54	2916	2916
53	3	159	2809	8427
52	2	104	2704	5408
51	4	204	2601	10404
50	1	50	2500	2500
49	3	147	2401	7203
48	2	96	2304	4608
47	2	94	2209	4418
46	1	46	2116	2116
45	2	90	2025	4050
44	1	44	1936	1936
43	1	43	1849	1849
41	1	41	1681	1681
38	1	38	1444	1444
37	1	37	1369	1369
33	1	33	1089	1089
	N = 34	1689 = (ΣX)		85395 = ΣX^2

$$\sigma^2 = \frac{\Sigma X^2 - (\Sigma X)^2/N}{N} = \frac{85395 - (1689)^2/34}{34}$$

$$= \frac{85395 - 2852721/34}{34}$$

$$= \frac{85395 - 83903.56}{34}$$

$$= \frac{1491.44}{34}$$

$$= 43.87$$

$\sigma = \sqrt{43.87} = 6.62$

NOTE: The values in column fX^2 may be obtained in two ways:

(a) $(f)(X^2)$ or (b) $(X)(fX)$

If $(X)(fX)$ is used you do not need the column X^2.

Go to frame 145.

145. The variance and standard deviation can be determined for distributions that have been grouped. The procedure is the same as above using the midpoints of the intervals for the X value. Use the tabled values from frame 32 to compute σ^2 and σ for the set of psychology test scores.

Interval midpoint X	f	fX	fX^2 $(X)(fX)$
64	1	64	4096
61	1	61	3721
58	2	116	6728
55	4	220	12100
52	9	468	24336
49	6	294	14406
46	5	230	10580
43	2	86	3698
40	1	40	1600
37	2	74	2738
34	1	34	1156
$N = 34$		$\Sigma X = 1687$	$\Sigma X^2 = 85159$

$N = 34 \quad (\Sigma X)^2/N = 83704.97 \quad \Sigma X^2 = 85159$

$$\sigma^2 = \frac{85159 - 83704.97}{34} = \frac{1454.03}{34} = 42.77$$

$$\sigma = \sqrt{42.77} = 6.54$$

Go to frame 146.

146. Since the variance and standard deviation give information of dispersion, they reflect differences among the scores.

What is the variance for the following set of scores? standard deviation?

37 37 37 37 37 37 37 37 37 37 37 37

0, 0 Right, if you are measuring differences and there are no differences the measure should be zero (0).

Go to frame 147.

Statistics of Variability

147. Which one of the following sets of measurements would be expected to provide the larger variance?

    A. set of times for nine-year old boys running 50 yards

    B. set of times for boys ages six to twelve running 50 yards

---

   B. Right! The greater the differences in the population the greater the variance (also standard deviation).
   A. No. Review the concept of variability.

   Go to frame 148.

148. Which of the following characteristics would be reflected in larger variances?

    A. heterogeneity   B. homogeneity

   A. Right!
   B. No. Populations that are alike will give small variances.

   Go to frame 149.

149. Given the following information, which group (A or B) would be the more homogeneous?

	Group A	Group B
$\bar{X}$	43.27	43.27
σ^2	17.64	9.48
σ	4.20	3.07
N	77	83

   B. Right!
   A. No. Although they have been measured with the same central location, the measures of variability reflect a greater scattering in group A, and less variation in group B.

   Go to frame 150.

150. (a) Which of the distributions below has the greatest variance?

    (b) Which has the largest standard deviation?

    (c) Which is leptokurtic?

    (d) Which is platykurtic?

    (e) Which represents a population that is the most homogeneous?

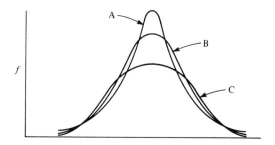

(a) C
(b) C
(c) A
(d) C
(e) A

Go to frame 151.

151. In the next section you will learn to use the standard deviation as a unit of measurement for deviation scores to build what is called a *standard score*. Standard scores are used much as other standard units of measure (inch, meter, liter, etc.) where a common unit is needed to make comparisons. Keep in mind as you work through the next chapter that a distribution of deviation scores $(X - \overline{X})$ has a mean of zero and a standard deviation of σ.

Go to frame 152.

5

STANDARD SCORES

152. In chapter 3 you learned to find a deviation score for each raw score and change the original distribution into a set of deviation scores. The procedure of changing a set of scores on certain characteristics but not others is called a *transformation*. A transformation is used for one of two reasons: it may help to meet the assumptions associated with a particular statistical procedure; but, more important to the area of descriptive statistics, it assists in the interpretation of scores. When Connie changed the test scores to deviation scores, she used a _____ .

---

transformation

Go to frame 153.

153. How did the transformation to deviation scores affect the rank order of the set of scores?

---

no change (ordinal property not affected)

Go to frame 154.

154. What characteristic of the frequency distribution is changed in the transformation $(X - \overline{X})$?

---

central location (mean)

Go to frame 155.

66 Standard Scores

155. How much are the scores changed?

They are reduced by the amount of the mean.

Go to frame 156.

156. Are any of the other characteristics (variability, skewness, kurtosis) affected?

No. All other characteristics remain the same.

Go to frame 157.

157. Explain in simple terms why the other characteristics are *not* affected.

The relationships within the set of scores remain the same. (See frame 140. The 13 and 16 differ by 3 units, and the corresponding deviation scores 1 and 4 differ by 3 units. All relationships are equal in the two distributions.) The graphic representations of the two distributions would also reveal no change, except the change of central location.

Here is the transformation pictorially:

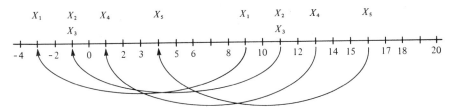

Go to frame 158.

158. Notice that any score that falls at the mean of the original distribution will have a deviation score of zero. If the above distribution had a score of 12 what would be its deviation score?

---

zero

Go to frame 159.

159. What is the mean of all sets of deviation scores?

---

zero

Go to frame 160.

160. What additional characteristic(s) will be affected if the deviation score is divided by the standard deviation? $(X - \bar{X})/\sigma$

---

variability

Go to frame 161.

161. How is each deviation score changed?

Each deviation score will now be expressed in units of the standard deviation.

Go to frame 162.

162. If a score in the original distribution is one standard deviation above the mean, the value after the two above transformations will be 1.

What new value will result from the two transformations for a score that is two standard deviations below the mean in the original distribution? three below? two above?

$^-2$, $^-3$, $^+2$

Go to frame 164.

164. If you combine the two transformations into one formula, $\dfrac{(X - \bar{X})}{\sigma}$, what is the mean and standard deviation of the new set of scores?

---

zero   1

Agree: Go to frame 165.
Disagree: Go to frame 152.

165. How is rank order affected by the transformation $(X - \bar{X})/\sigma$? skewness?   kurtosis?

---

no change, no change, no change

Go to frame 166.

68 Standard Scores

166. The transformation $(X - \bar{X})/\sigma$ is used to put obtained scores into *standard score* form. The standard score (z) is used as a standard measure much as other standard measures (inch, liter, etc.) are used. Any standard measure uses a standard unit. Using the z-score, comparisons between two distributions can be made directly because with z-scores the two distributions have the same mean and the same standard deviation. Students' test scores recorded as standard scores are more easily interpreted and compared. Differences and trends in scores become more apparent through the use of standard scores since values in different distributions are directly comparable. What formula is used to find a standard score (z)?

$$z = \frac{X - \bar{X}}{\sigma}$$

Note: z-scores are reported to the nearest hundredth.
Examples: +1.00, -1.32, -2.15

Go to frame 167.

167. Determine z-scores for the following scores on the psychology test:
44, 52, 56, 61

$(44 - 49.68)/6.62 = -5.68/6.62 = -.86$
$(52 - 49.68)/6.62 = +2.32/6.62 = +.35$
$(56 - 49.68)/6.62 = +6.32/6.62 = +.95$
$(61 - 49.68)/6.62 = +11.32/6.62 = +1.71$

Go to frame 168.

168. What is the z-score for 65 (highest score)?
What is the z-score for 33 (lowest score)?

$z_{65} = (65 - 49.68)/6.62 = +15.32/6.62 = +2.31$
$z_{33} = (33 - 49.68)/6.62 = -16.68/6.62 = -2.52$

Go to frame 169.

169. Notice from the two values in frame 168 that what was a large range (65 - 33) is now relatively small. This fact and the involvement of decimals and negative numbers in z-score values are all disadvantages of z-score distributions. Another transformation *overcomes the disadvantages* while *keeping the advantages* of a standard score. This transformation is commonly referred to as the Z-score transformation. Z-scores are found by multiplying each z-score by 10 and adding 50. $Z = 10(z) + 50$

What Z value does a z-score of zero take?

50

Go to frame 170.

170. What is the mean of the Z-score distribution?

    What is the standard deviation of the Z-score distribution?

$\overline{X} = 50$ $\sigma = 10$

Agree: Go to frame 172.
Disagree: Go to frame 171.

171. The Z- score formula has the effect of adding 50 points to each z-score value much the same as taking away when we subtracted a mean value from all scores in finding deviation scores. Notice that this affects only the characteristic of central location and overcomes the disadvantage of some values being negative and some values being positive.

    The Z-score formula also has the effect of enlarging differences (in fact differences are increased 10 times). By rounding to whole numbers (after multiplication) the decimals are eliminated.

    Determine Z-scores for $z = 0.00$, $z = +1.00$, $z = +2.00$, $z = +3.00$, $z = -3.00$, $z = -2.00$, $z = -1.00$.

50, 60, 70, 80, 20, 30, 40

Go to frame 172.

172. Determine Z-scores for the following psychology test scores: 33, 44, 52, 56, 61, 65 (See frames 167 and 168 for z-scores.)

25, 41, 54, 60, 67, 73

Go to frame 173.

173. The z-score and Z-score are two examples of *standard scores*. Any desired mean and standard deviation can be built into a standard score distribution by manipulating two constants in the formula:

$$\text{standard score} = \underline{d}\ (z) + \underline{c}$$

where the constant $\underline{c}$ becomes the mean, and the constant $\underline{d}$ becomes the standard deviation.

Standardization of widely used tests has resulted in many variations of the standard score. Some examples are:

$$\begin{aligned}
\text{CEEB scores} &= 100(z) + 500 \\
\text{AGCT scores} &= 20(z) + 100 \\
\text{Stanford-Binet scores} &= 16(z) + 100
\end{aligned}$$

A score of 650 on the CEEB test is _____ standard deviations (above, below) the mean.

---

one and one-half, above

Agree: Go to frame 175.
Disagree: Go to frame 174.

174. 650 is 150 units above 500. (Deviation score = +150) Putting 150 units into standard deviation units gives 150/100 = +1.50.

The positive sign tells us that it is above the mean, 1.50 indicates one and one-half standard deviation units.

Go to frame 173.

175. A popular standard score used widely especially in the area of testing is the T-score. The T-score like the Z-score has a distribution mean of 50 and a standard deviation of 10. The only difference between the two distributions is that the T-score distribution has been *normalized* by converting the scores so that any skewness in the distribution disappears. The next chapter, "The Normal Curve," will discuss the properties of the normal curve. The conversion procedure is presented in many introductory statistics books and would be valuable for use with heavily skewed distributions or published standardized tests.[1]

Go to frame 176.

---

1. George A. Ferguson, *Statistical Analysis in Psychology and Education*. 3rd ed. (New York: McGraw-Hill, 1971), pp. 383-86.

6

THE NORMAL CURVE

177. There is a tendency for measurements of most variables to cluster around some central location. Investigation of sets of data in tabular or graphic form point up this phenomenon. It is more apparent with distributions with large Ns, but with very small numbers it may not be obvious. With an N equal to 5 it is not likely that this tendency would be exhibited, but as the frequencies increase the tendency to cluster will become more and more apparent. This happens because most of the variables that the behavioral sciences deal with are "normally distributed." If frequency distributions of natural events were graphed the frequency curve would usually take the shape of a bell. In fact, the term "bell-shaped" is used to describe a frequency distribution that approaches the shape of a family of curves that have common characteristics. Because of this tendency the curve that is generated with a mean of zero and a standard deviation of one is used as a model. We will refer to this special curve as the *normal curve* since it is most appropriate to help interpret scores.

What is the name of the curve that has $\overline{X} = 0$, and $\sigma = 1$, and is used as a model?

---

normal curve (the one in standard form)

Go to frame 178.

72 The Normal Curve

178. There is an equation (based on the mean and variance values) that generates a normal curve, but it will not be given here. (See an introductory statistics book for the formula and its use.[1])

Keep in mind:

1. the equation would give many curves that would be considered normal because of common properties;
2. we have chosen the one in standard form for our model;
3. the normal curve is built from every conceivable value on the measured variable.

The normal curve is (a) or (b).

(a) continuous, or
(b) discrete

(a) continuous

Go to frame 179.

179. Because all values must be considered, the two "tails" of a frequency polygon for a normal curve would never touch the horizontal axis. The tails of the graph will be *asymptotic* to the abscissa.

Graphically the tails of the normal curve will approach but never _____ the abscissa.

---

touch (reach)

Go to frame 180.

180. In a normal curve the mean, median, and mode take the same value.

What is the *median value* for the normal curve model? mode?

---

zero   zero

Go to frame 181.

---

1. George A. Ferguson, *Statistical Analysis in Psychology and Education*. 3rd ed. (New York: McGraw-Hill, 1971), p. 88.

The Normal Curve 73

181. The normal curve is symmetrical.

How is the normal curve skewed?

---

no skewness (symmetrical) (The frequency polygon divided at the mean would produce two halves that are the same shape.)

Go to frame 182.

182. The normal curve is neither leptokurtic nor platykurtic. It has a medium degree of kurtosis.

How would you describe the characteristic of kurtosis for the normal curve?

---

mesokurtic

Go to frame 183.

For frames 183 to 186 choose the frequency curve that could NOT be a normal curve, and tell why it could not be.

183.

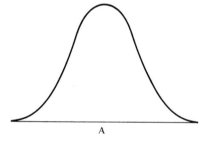 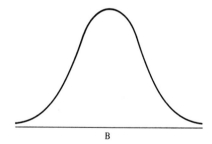

A. tails touch the base line

184.

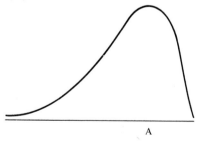 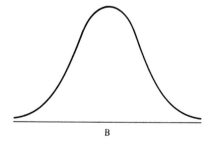

A. negatively skewed

74 The Normal Curve

185.

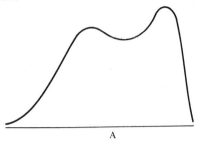
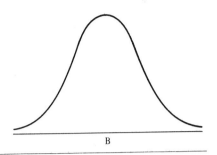

A. bimodal (the normal curve has the same value for mean, median, and mode)

186.

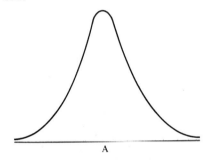

A. too leptokurtic (normal curve is mesokurtic)

Go to frame 187.

187. The bell-shaped curve below approximates a normal curve. Review its properties:

    1. Tails of the curve are asymptotic to the abscissa
    2. Symmetrical
    3. Mean, median, and mode have the same value
    4. Mesokurtic

Also observe:

    5. The curve changes direction (inflection point) at the points that are one standard deviation on either side of the mean.
    6. The maximum ordinate value is at the mean.

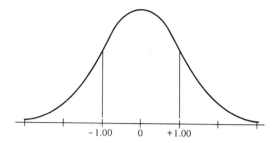

Go to frame 188.

188. A normal curve with a mean of zero and a standard deviation of one represents a normal distribution of standard scores. Nearly all of the area under the curve will fall between the point that is three standard deviations to the left of the mean and the point that is three standard deviations to the right. In the normal curve, what proportion of the area of the curve falls above the mean? below the mean?

50%   50%

Go to frame 189.

189. The true representation of the normal curve can not be drawn since the tails of the curve never touch the base line and would continue indefinitely in each direction. However, the graph below and the associated percentages represent roughly the proportions of the area for chosen segments of the base line. Tables are available to permit you to make even more precise estimates and provide accurate percentages for base line segments using the mean for a reference point. It is customary to give proportions for each z-score value for each hundredths place.

The table would provide these rough percentages for areas under the curve.

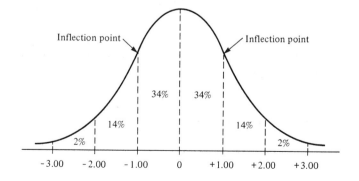

76 The Normal Curve

The area under the curve is considered unity (1) and is tabled in proportion to the total area. The tabled value for the line segment (abscissa) from the mean to one standard deviation above the mean is .3413. Changing this from a proportion to a percentage, about 34% of the area under the normal curve falls within the shaded section of the graph below.

What proportion would be associated with the unshaded area above the mean?

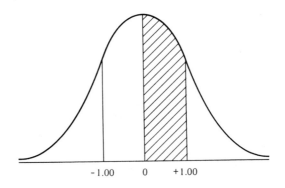

16%

Agree: Go to frame 191.
Disagree: Go to frame 190.

190. The perpendicular line drawn from the mean divides the area into two equal parts (50% below and 50% above). If 34% is between the mean and +1.00 then 16% must be above the point +1.00.

Go to frame 189.

191. What part of the area is below the value +1.00?

84% (50% below mean plus 34% to +1.00)

Go to frame 192.

192. About what part of the area of the curve is associated with the segment from -1.00 to +1.00?

---

roughly 2/3 (68.26% using the tabled values)

Go to frame 193.

193. The area under a frequency polygon or histogram represents the total frequency of the distribution.

The area under a normal curve represents the total frequency of a set of normally distributed standard scores. If a set of obtained scores can be considered as normally distributed, then the normal curve and its properties can be used to interpret relationships within a set of obtained scores.

(a) *What part* of 600 normally distributed scores would be expected to lie between the two points -1.00 and +1.00?

(b) *How many* of the 600 scores in standard score form would take values between -1.00 and +1.00?

(a) about 2/3 (68.26%)
(b) about 400   2/3 of 600 = 400
    409.56   68.26 · 600 = 409.56
    NOTE: Since the set of obtained scores only approximates the normal curve, our estimates can be made only roughly. It is not likely (impossible) for exactly 409.65 observations to take values between -1.00 and +1.00. 410 might be a better estimate than 400—you decide.

Agree: Go to frame 195.
Disagree: Go to frame 194.

194. Question 193(a) is basically the same question asked in frame 192. Frame 189 shows 34% in each of the two middle areas. The total of these two areas would represent the proportion of the curve associated with the segment from -1.00 to +1.00.

Question 193(b) asks that you put this proportion of the curve area into a number of cases. In this particular problem the area under all of the curve represents 600 cases. The question becomes what is 2/3 of 600 or more precisely what is 68.26% of 600?

Go to frame 193.

78 The Normal Curve

195. (a) *What part* of 600 normally distributed scores would be expected to lie between the two points −2.00 and +2.00? between −3.00 and +3.00?

(b) *How many* of the 600 scores in standard score form would take values between −2.00 and +2.00? between −3.00 and +3.00?

(a) 96% (table 95.44%)
100% (table 99.74%)
(b) 576 (table 572.64)
600 (table 598.44) (For the 600 about one or two would be expected to score above +3.00 or below −3.00)
(For 10,000 cases about 26 would fall beyond these limits.)

Go to frame 196.

196. (a) *What part* of the distribution of psychology scores would be expected to fall below a score of 52?

(b) *How many* of the 34 students would be expected to score below 52?

$z_{52} = +.35$

Tabled values

z-score	area
.32	.1255
.33	.1293
.34	.1331
.35	.1368
.36	.1406

(a) .5000 + .1368 = .6368 or about 64%
(b) 22

Agree: Go to frame 198.
Disagree: Go to frame 197.

197. The z-score places the raw score in relation to the mean—direction and distance. The tabled value gives the proportion of the curve between the mean and the z-score. The z-score for a raw score of 52 is +.35 (see frame 167). The tabled value for .35 is .1368 (see frame 196). Pictorially:

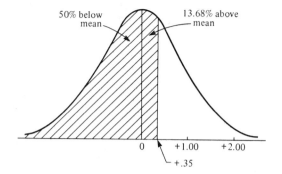

63.68% of the curve falls below +.35
63.68% of the frequencies (34) fall below +.35
63.68% of 34 = 21.65

Go to frame 196.

198. How many of the 34 students would be expected to score below 44? How many below 56?

$z_{44} = -.86$
$z_{56} = +.95$

Tabled values

z-score	area
.85	.3023
.86	.3051
.87	.3078
.88	.3106
.89	.3133
.90	.3159
.91	.3186
.92	.3212
.93	.3238
.94	.3264
.95	.3289
.96	.3315

about 7 (6.63) below 44
about 28 (28.18) below 56

Agree: Go to frame 200.
Disagree: Go to frame 199.

80 The Normal Curve

199. 50% + 32.89% = 82.89% will fall below 56
82.89 of 34 = 28.1826

50% - 30.51% = 19.49% will fall below 44
19.49% of 34 = 6.6266

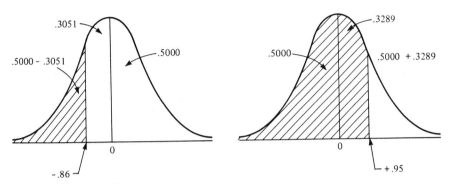

Go to frame 198.

200. Professor Ego gave a second test to his psychology 201 class. Connie was interested in knowing how the two students who scored 52 on the first test did on this test. She was disappointed to find that each of them scored lower on the second test. One scored 48 and the other scored 50.

The mean for the second test was 47.82 and the standard deviation was 5.46.

Do you agree with Connie that both students' scores on the second test were poorer than the score of 52 on the first test?

In relation to the performance of the class the score of 48 is a lower score, but the 50 is a higher score.

Agree: Go to frame 202.
Disagree: Go to frame 201.

201. It is impossible to make a valid comparison of raw scores on two different tests in the form of raw scores. The standard score (z) can be used as the common unit to aid in making the comparison. On the first test each student who scored 52 was in the same place in relation to others being tested. Each scored a z-score of +.35 which put them .35 of one standard deviation above the mean. On the second test the relative positions are:

$$z_{48} = \frac{48 - 47.82}{5.46} = +.03 \quad Z = 50$$

$$z_{50} = \frac{50 - 47.82}{5.46} = +.40 \quad Z = 54$$

The student with a 48 on the second test has moved down in the distribution, while the student with a 50 has moved (slightly) up in the second distribution of test scores.

Go to frame 202.

202. The use of the standard score includes:

    1. comparison of performance within a group;
    2. comparison of an individual's performance in two or more situations;
    3. comparison of an individual's performance to some norming group.

In each of these, characteristics must be known to make meaningful interpretations. There must be some basis for comparison.

Comparing performance of an individual to others in a group requires that certain important characteristics be the same for all. Age, I.Q., socioeconomic background, and/or others may be important in making valid comparisons. To view trends or changes in individual performance the same group should be used. In comparing the two test scores for the psychology tests, the scores were obtained from the same class. If one of the students who scored 52 on the first test had changed classes and/or instructor, any differences would have been clouded by changing the variables and their contributions to the differences. It is important to keep everything (variables) the same (constant) when making comparisons.

When interpreting a score to norms it is important to know the characteristics of the norm group. A meaningful comparison can be made only if the individual's characteristics fit the norming group.

Go to frame 203.

7

CORRELATION

203. The interpretation of sets of data and individual values has been largely a study of relationships. For example, we viewed scores as they related to the mean, and developed a standard score that helped us to study the relationship of one score in one distribution with a score in another distribution. Although two distributions were involved we considered only one variable. Data of this nature are referred to as *univariate data*.

Many questions that arise in testing and research involve two variables. In this chapter we turn to relationships that exist between two sets of data and the simultaneous variation of the two variables. A study of the degree of relationship between two variables calls for a correlational technique. *Correlation* is a study of the relationship that exists between two variables.

Data gathered on one variable are called univariate data and data consisting of pairs of measurements used to study relationships are called _____.

---

bivariate data

Go to frame 204.

204. Examples of *questions* that arise involving relationships are:

    Do these two tests measure the same thing?
    Are these two tests equivalent?
    Do each of these two tests measure achievement in high school biology to the same degree?

84 Correlation

How well will scores on this test predict a student's future performance in college?

What is the relationship of reading comprehension and achievement in United States history?

What is the relationship between hours spent in extra-curricular activities and grade point average in high school?

Notice that each of the above questions involves two variables. Gathering data for answering a relationship question requires two pieces of information on common subjects. Correlation deals with paired measurements (one on each of the two variables) on each common element.

For each of the following pairs of variables write a question that would require a correlational technique to answer.

(a) height of seventh-grade students
weight of seventh-grade students

(b) participation in athletics
scholastic achievement

(c) speed of an automobile
gasoline mileage

(d) ratings of pieces of art by
(1) college art majors
(2) elementary school students

---

correct answers will vary

Some examples of correct and incorrect answers.

(a) CORRECT: What is the relationship between height and weight for seventh-grade students?

INCORRECT: Is there a correlation between height and weight for seventh-grade students? [The answer to the incorrect question is "yes." A correlation can be obtained for any two sets of data (values) on common subjects.]

(b) CORRECT: How are the two variables—athletic participation and scholastic achievement—related for male students in the College of Arts and Sciences?

INCORRECT: Does participation in athletics cause lower grades for men students in the College of Arts and Sciences? (Correlation is a study of relationships. It says nothing about *causes*. Cause and effect relationships can be hypothesized from derived correlations, but correlational procedures do not identify causes.)

(c) CORRECT: How are high speeds of an automobile associated with gasoline mileage as compared to gasoline mileage for slower speeds?

INCORRECT: Do high speeds cause low gasoline mileage? (Key word is "cause." If a correlation revealed an association of low gas mileage with high speeds then a hypothesis could be developed. One possible hypothesis: Gas mileage is reduced at high rates of speed because of the reduced engine efficiency.)

(d) CORRECT: How do ratings of art objects by elementary school children compare to ratings given by college art majors?

INCORRECT: Does lack of training in art cause elementary school students to rate art objects of high quality poorly?

Go to frame 205.

205. Professor Ego conducted a seminar for senior psychology students. At the end of the term each of the eight class members submitted a term paper as part of the course requirements. To aid in the evaluation of the papers he made a class ranking of the papers. His ranking was:

Ann	1
Ben	2
Carl	3
Dan	4
Eileen	5
Fred	6
Gayla	7
Howard	8

Professor Ego asked Connie and Professor Idson to rank the papers.

Connie's ranking		Professor Idson's ranking	
Ann	1	Ann	2
Ben	3	Ben	3
Carl	2	Carl	1
Dan	5	Dan	4
Eileen	4	Eileen	5
Fred	6	Fred	6
Gayla	7	Gayla	8
Howard	8	Howard	7

Did any two of the rankings give the same rank to each student?

---

no

Go to frame 206.

86 Correlation

206. Although the rankings were not the same there appears to be a common trend between any two of the rankings. The students who had high ranks in one ranking tended to have high ranks in the others—middle ranks tended to remain in the middle, and low ranks on one ranking tended to be low in the others. Any trend within data is referred to as a relationship. This relationship is measured statistically by *correlation*. If there had been no trend within the data the correlation would reveal that there is no relationship. A correlation of near 0 .00 would indicate no relationship. If there is little or no relationship within the two sets of data being correlated, the correlation coefficient will be about _____ .

0.00

Go to frame 207.

207. Another way to look for relationships is by a *scattergram*. Relationships may be examined by plotting points on a graph and letting each pair of observations (the two measures on each subject) be represented by a point. Let each of the two dimensions represent values on one variable and plot the paired values with one point. The following scattergram (scatter diagram) reveals no discernible relationship between the two variables. The dotted lines for one point (A) indicate how the scattergram is built. Every other point represents two measurements for a subject.

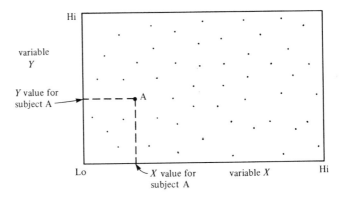

There does not appear to be a trend within the scattergram nor a relationship between the variables. The correlation coefficient for this scattergram would be near _____ .

0.00

Go to frame 208.

208. If low scores on one variable are associated with low scores on the other variable, middle with middle, and high with high, the scattergram would tend to take this general form.

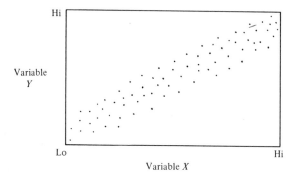

The closer the dots are to forming a line the higher the correlation. If they fall on a line with low-low to high-high relationship, the coefficient

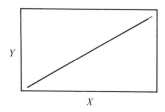

will be equal to +1.00. This is referred to as *perfect correlation*. A perfect low-low to high-high relationship will generate a correlation coefficient equal to _____.

+1.00

Go to frame 209.

88 Correlation

209. If low scores on one variable are associated with high scores on the other variable and high with low, the scattergram will look like this.

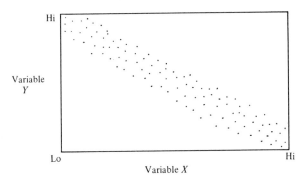

The closer the dots are to forming a line the higher the correlation. If they fall on a line with low-high and high-low relationship the coefficient

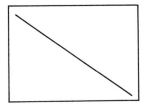

will be equal to -1.00. This is referred to as *perfect correlation*. A perfect low-high to high-low relationship will generate a correlation coefficient equal to _____.

-1.00

Go to frame 210.

210.

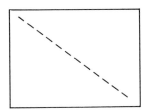  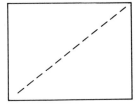

(a) Which of the above scattergrams would indicate perfect relationship?

(b) Which indicates no discernible relationship?

(a) A and C
(b) B

Agree: Go to frame 212.
Disagree: Go to frame 211.

211. When the plotted points in a scattergram fall on a line the relationship is at a maximum. Both *A* and *C* satisfy this. However, *there is one difference.* The direction of the line in *A* is from high to low (moving to the right) and the direction of *C* is from low to high (moving to the right). In either case, there is perfect correlation. Since B reveals no trend (pattern), there is no relationship.

Go to frame 210.

212. The correlation coefficient can be thought of as having two parts: (1) the size of the coefficient that indicates the degree (amount or extent) of relationship; and (2) the sign that gives the direction.

   (a) What is the degree of relationship for scattergrams *A* and *C*?

   (b) What is the correlation coefficient for *A*? for *C*?

(a) perfect relationship
(b) -1.00, +1.00

Agree: Go to frame 213.
Disagree: Go to frames 208 and 209 for review.

213. (a) How do the two values (-1.00 for *A* and +1.00 for *C*) differ?

   (b) How do you interpret this difference?

(a) the signs (coefficient same size but *signs differ*)
(b) relationships are in different directions (same degree of relationship but different directions)

Agree: Go to frame 214.
Disagree: Go to frame 211.

214. *If someone had ranked the eight term papers* from Professor Ego's seminar exactly as he did, the rankings would be as follows:

	Professor Ego	Other
Ann	1	1
Ben	2	2
Carl	3	3
Dan	4	4
Eileen	5	5
Fred	6	6
Gayla	7	7
Howard	8	8

This is a scattergram for these two rankings:

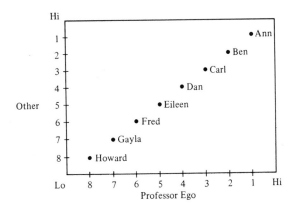

The correlation coefficient for these two sets of rankings would be

_____ .

What degree of relationship? What direction?

+1.00
perfect relationship   positive direction

Go to frame 215.

215. If someone had ranked the eight term papers from Professor Ego's seminar class in inverse order, the rankings would be as follows:

	Professor Ego	Other
Ann	1	8
Ben	2	7
Carl	3	6
Dan	4	5
Eileen	5	4
Fred	6	3
Gayla	7	2
Howard	8	1

This is a scattergram for these two rankings:

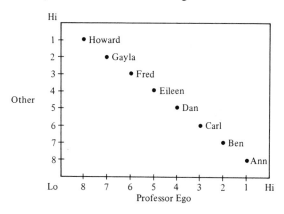

The correlation coefficient for these two sets of rankings would be _____.

What degree of relationship? What direction?

−1.00
perfect relationship   negative direction

Go to frame 216.

92 Correlation

216. The formula for finding the correlation coefficient with data that are ranked is as follows:

$$\rho \ (rho) = 1 - \frac{6\Sigma D^2}{N(N^2 - 1)}, \text{ where,}$$

ρ is the correlation coefficient
D is the difference in ranks (each subject)
ΣD^2 is the sum of squared differences
N is the number of pairs

Determine ρ for the data from frame 214.

+1.00

Agree: Go to frame 218.
Disagree: Go to frame 217.

217.

Student	Ranks		Differences	
			D	D^2
Ann	1	1	0	0
Ben	2	2	0	0
Carl	3	3	0	0
Dan	4	4	0	0
Eileen	5	5	0	0
Fred	6	6	0	0
Gayla	7	7	0	0
Howard	8	8	0	0
				$0 = \Sigma D^2$

$$\rho = 1 - \frac{6(0)}{8(64-1)} = 1 - \frac{0}{504} = 1 - 0 = +1.00$$

Go to frame 218.

218. Determine *rho* for the data from frame 215.

-1.00

Student	Ranks		Differences	
Ann	1	8	-7	49
Ben	2	7	-5	25
Carl	3	6	-3	9
Dan	4	5	-1	1
Eileen	5	4	1	1
Fred	6	3	3	9
Gayla	7	2	5	25
Howard	8	1	7	49

$$168 = \Sigma D^2$$

$$\rho = 1 - \frac{6(168)}{8(64-1)} = 1 - \frac{1008}{504} = 1 - 2 = -1.00$$

Go to frame 219.

219. Determine *rho* for the following pairs of rankings (see frame 205):
   A. Professor Ego-Connie Coefficient
   B. Professor Ego-Professor Idson
   C. Professor Idson-Connie Coefficient

A					B					C			
1	1	0	0		1	2	-1	1		1	2	-1	1
2	3	-1	1		2	3	-1	1		3	3	0	0
3	2	1	1		3	1	2	4		2	1	1	1
4	5	-1	1		4	4	0	0		5	4	1	1
5	4	1	1		5	5	0	0		4	5	-1	1
6	6	0	0		6	6	0	0		6	6	0	0
7	7	0	0		7	8	-1	1		7	8	-1	1
8	8	0	0		8	7	1	1		8	7	1	1
			4					8					6

$\rho = 1 - \frac{24}{504}$ $\rho = 1 - \frac{48}{504}$ $\rho = 1 - \frac{36}{504}$
$= 1 - .05$ $= 1 - .10$ $= 1 - .07$
$= +.95$ $= +.90$ $= +.93$

Go to frame 220.

94 Correlation

220. The above method of correlation (Spearman Rank Difference) is one of several correlational techniques and can be used with ranked data or sets of data that can be ranked. The closer that the value of *rho* is to +1.00 or -1.00 the greater the relationship, and the closer to 0.00 the less the relationship. The three correlation coefficients derived from Professors Ego and Idson and Connie rankings revealed a high relationship each time.

Would this indicate that there was a high or low degree of agreement on quality of the papers?

How would disagreement be reflected by coefficient values?

---

high degree of agreement (high positive relationship)
by a negative relationship (see frame 215)

Go to frame 221.

221. Another correlational procedure that utilizes obtained measurement scores is called the Pearson product-moment method. The basic formula in standard score form is $r = \frac{\Sigma(z_x z_y)}{N}$. If the pairs of z-scores take the same value (including sign) the resulting r will equal +1.00. If the z-scores are the same but the signs of each pair are different, r will equal -1.00. As before, the limits for r become -1.00 and +1.00.

How do you think a lack of any relationship would be indicated by r values?

0.00 (same as with *rho*)

Go to frame 222.

222. A more convenient formula (an identity) for hand computation utilizes deviations to shorten the amount of arithmetic and should be used rather than the above formula. The more efficient formula is:

$$r = \frac{\Sigma xy}{\sqrt{(\Sigma x^2)(\Sigma y^2)}}, \text{ where,}$$

$x = X - \bar{X}$
$y = Y - \bar{Y}$
x^2 = deviation score (x) squared
y^2 = deviation score (y) squared
xy = product of the paired deviation scores
Σ means to sum

This is an example of calculation of r using deviation scores:

Student	Raw scores		$X - \bar{X}$	$Y - \bar{Y}$	$(X - \bar{X})^2$	$(Y - \bar{Y})^2$	
	X	Y	x	y	x^2	y^2	xy
Ian	14	7	4	2	16	4	8
John	15	5	5	0	25	0	0
Kim	11	6	1	1	1	1	1
Les	10	7	0	2	0	4	0
Mae	10	4	0	-1	0	1	0
Nan	9	7	-1	2	1	4	-2
Opal	5	2	-5	-3	25	9	15
Pete	6	2	-4	-3	16	9	12
	80	40	0	0	84	32	34

$\bar{X} = 10.00 \quad \bar{Y} = 5.00 \qquad \Sigma x^2 = 84 \quad \Sigma y^2 = 32 \quad \Sigma xy = 34$

$$r = \frac{\Sigma xy}{\sqrt{(\Sigma x^2)(\Sigma y^2)}} = \frac{34}{\sqrt{(84)(32)}} = \frac{34}{51.84} = +.66$$

If a calculator or adding machine is available, the following formula will be even quicker. The only change from the deviation score formula is substituting identities for Σxy, Σx^2, Σy^2.

$$r = \frac{\Sigma XY - \frac{(\Sigma X)(\Sigma Y)}{N}}{\sqrt{\left[\Sigma X^2 - \frac{(\Sigma X)^2}{N}\right]\left[\Sigma Y^2 - \frac{(\Sigma Y)^2}{N}\right]}}, \text{ where,}$$

X is a raw score on the abscissa variable.
Y is a raw score on the ordinate variable.
N is the number of paired measurements.

Using the above data, compute Pearson r utilizing the raw score formula.

$r = +.66$

See frame 223 for computation

96 Correlation

223.

X	Y	X^2	Y^2	XY
14	7	196	49	98
15	5	225	25	75
11	6	121	36	66
10	7	100	49	70
10	4	100	16	40
9	7	81	49	63
5	2	25	4	10
6	2	36	4	12
80	40	884	232	434
ΣX	ΣY	ΣX^2	ΣY^2	ΣXY

$$r = \frac{434 - \frac{(80)(40)}{8}}{\sqrt{\left[884 - \frac{(80)^2}{8}\right]\left[232 - \frac{(40)^2}{8}\right]}}$$

$$= \frac{434 - 400}{\sqrt{(884 - 800)(232 - 200)}}$$

$$= \frac{34}{\sqrt{(84)(32)}} = \frac{34}{51.84} = +.66$$

Before going on, review both procedures for determining the Pearson r. (Notice the common steps in calculation—last three.)

Go to frame 224.

224. Using the data from frame 222, rank the scores in the two distributions and calculate *rho*.

Student	Scores		Ranks		Differences	
	X	Y	X	Y	D	D^2
Ian	14	7	2	2	0	0
John	15	5	1	5	-4	16
Kim	11	6	3	4	-1	1
Les	10	7	4.5	2	2.5	6.25
Mae	10	4	4.5	6	-1.5	2.25
Nan	9	7	6	2	4	16
Opal	5	2	8	7.5	.5	.25
Pete	6	2	7	7.5	-.5	.25
						42.00

$$\rho = 1 - \frac{6(42)}{8(64-1)} = 1 - \frac{252}{504} = 1 - .50 = +.50$$

The relatively large difference between r and *rho* for the set of data is caused by the small number of subjects and the fact that a rather large proportion of the scores were ties. The derivation of the formula for finding *rho* assumes no ties. If there are relatively few ties the *rho* value is a good estimate, but as you can see here ties have affected our value rather severely.

Go to frame 225.

Correlation

225. Interpretation of *r* values of −1.00, 0.00, and +1.00 is straightforward and we have found that they represent, respectively, perfect negative relationship, no linear relationship, and perfect positive relationship.

Interpretation of values such as +.66 takes a different turn. It is *not* correct to say that there is a 66% relationship. There seems to be a tendency to want to interpret *r* values as percentages because the coefficients look much like a proportion or a percentage. The correlation coefficient is *not* a direct measure of the percentage of relationship between the two variables. Since the study of correlation is a study of shared variances the correlation coefficient utilizes a two-dimensional statistic, the variance. A more comprehensive study of correlation reveals that you can speak in terms of percentage of relationship (shared variance) by squaring the *r* value and multiplying by 100.

What degree of relationship is measured by the following correlation coefficients?

+.71   +.50   −.71   −.50   −1.00   +1.00   0.00   +.66

---

about 50%, 25%, about 50%, 25%, 100%, 100%, 0%, 44%

Agree: Go to frame 226.
Disagree: Use the formula ($r^2 \cdot 100$)
   Example:
   $(+.71)(+.71) \cdot 100 = .5041 \cdot 100 = .5041$ or 50%

Go to frame 226.

226. Some *r* values with the corresponding relationships

±*r*	$r^2 \cdot 100$
.10	1
.20	4
.30	9
.40	16
.50	25
.60	36
.70	49
.80	64
.90	81
1.00	100

NOTICE: It takes a value of +.70 or −.70 to approach a 50% shared variance, and even ±.90 leaves 19% of the variance unaccounted for.

98 Correlation

Since the correlation coefficients are not on a scale with equal units, intervals within the scale are not the same, for example, the distance from +.20 to +.60 is *not* the same as from +.50 to +.90 although subtraction of +.20 from +.60 gives the same answer as +.50 from +.90. To move from +.20 to +.60 or from +.50 to +.90 reveals an *increase* in the *measure of relationship* but *not* in the same _____.

---

amount (degree)

Go to frame 227.

227. The interpretation of r values involves the questions: "What r is required for a high relationship?" and "What r is considered to be a low relationship?" Other than the three points -1.00, 0.00, and +1.00 there is no one clear cut answer. The situation being investigated and the variables under consideration will cause the values to vary. Each situation must be interpreted with its own uniqueness. When establishing the reliability or equivalence of standardized achievement tests, $r = +.70$ might be considered a very low relationship, but be quite acceptable for an informal classroom test built to be used only one time. The standardized test maker might require $r = +.90$ for his standardized achievement test while his colleague developing a test for predicting future performance from one test administration might be fortunate to get $r = +.75$. When deciding what an r value means, it is well to see what others have been able to obtain using the same two variables.

Certainly .90 to 1.00 (either negative or positive) shows high relationship and 0.00 to .30 (either - or +) shows low relationship—the interpretation of the rest is in terms of the variables.

Go on to the last frame—228.

228. The author of the material you have been using to study procedures to work with sets of data hopes that it will have utility for your needs and that you feel that you want to find out more about these procedures and others available to you to answer other types of questions. These answers may not be as clear cut and direct as you (and I) would like, since much interpretation must be used to make decisions about what answers are in the messages found in sets of data. (See frame 227.) Statistical procedures provide ways to help make valid decisions about important problems and answers to important questions. The more you use them the more skilled you will become in their utilization in solving your problems. Good luck.

Appendix 1

COMPUTATIONAL MASTERY TEST

1. The following set of scores was obtained from fourteen students learning the addition facts. The values represent the numbers of errors made on a trial of fifty selected facts.

    17 7 3 12 13 14 12
    7 14 5 20 7 9 11

    (a) Order the scores.
    (b) Rank the scores.
    (c) What is the rank for a score of 17? 7? 11?

2. The scores on p. 100 were obtained from fifty students (two classes) who were being measured on their achievement in mathematics. The values represent the number of correct answers for fifteen multiplication problems for one-digit multipliers.

Example of the task: 3 · 14 = ☐

Possible scores	Frequency (f)
15	0
14	1
13	5
12	7
11	12
10	10
9	5
8	3
7	2
6	2
5	1
4	1
3	1
2	0
1	0
0	0

Using these data provide the information requested below.

(a) N
(b) Range
(c) Mean ($\bar{X}$)
(d) Median
(e) Mode
(f) Variance (σ^2)
(g) Standard deviation (σ)
(h) Third quartile (Q_3)
(i) Thirty-fifth percentile (P_{35})
(j) Sixth decile (D_6)
(k) z-scores for scores 3 through 14
(l) Z-scores for scores 3 through 14
(m) Skewness (positive or negative)

3. Preference for ten art objects was obtained from two judges. Their rankings are as follows:

Object	Judge A	Judge B
A	3	1
B	10	9
C	7	7
D	2	3
E	1	2
F	9	10
G	8	8
H	5	4
I	6	5
J	4	6

(a) Prepare a scattergram to investigate for relationship.
(b) Is there a discernible relationship?
   If yes, is the relationship positive or negative?
(c) Use the formula for finding *rho* to obtain a correlation coefficient for the data.

$$\rho = 1 - \frac{6\Sigma D^2}{N(N^2 - 1)}$$

4. Obtain the Pearson *r* for the provided data.

(a) Use the deviation formula

$$r = \frac{\Sigma xy}{\sqrt{(\Sigma x^2)(\Sigma y^2)}}$$

(b) Use the raw score formula

$$r = \frac{\Sigma XY - \Sigma X \Sigma Y/N}{\sqrt{(\Sigma X^2 - \Sigma X/N)(\Sigma Y^2 - \Sigma Y/N)}}$$

Appendix 1

	Raw Scores								
	X	Y	x	y	x^2	y^2	xy	X^2	XY
I	10	12							
J	14	16							
K	3	3							
L	8	11							
M	7	6							
N	2	5							
O	12	14							
P	8	13							
SUMS									

ANSWERS FOR COMPUTATIONAL MASTERY TEST

Frames
for
review

1 and 2   1. (a) ordering
                3 5 7 7 7 9 11 12 12 13 14 14 17 20

3 to 8   (b) ranking
                1 2 4 4 4 6 7 8.5 8.5 10 11.5 11.5 13 14

The score of three was assigned a one because it represented the highest score (fewest errors). The ranking could start by assigning the one to an error score of 20. The ranking then would be

ordering
3 5 7 7 7 9 11 12 12 13 14 14 17 20
ranking
14 13 11 11 11 9 8 6.5 6.5 5 3.5 3.5 2 1

7   (c) 13, 4, 7

13   2. (a) N = 50
15 and 132   (b) Range = 14 − 3 = 11
79 to 82   (c) $\bar{X}$ = 10.04
113   (d) Median = 10.50
58   (e) Mode = 11
134 to 137   (f) Variance = 5.56
and 141 σ^2 = 5.56
139 to 140   (g) St. Dev. = 2.36
and 141 σ = 2.36
106 and 107 (h) Q_3 = 11.57
110 to 113   (i) P_{35} = 9.75
108 to 109   (j) D_6 = 10.92
166 to 168   (k) and (1)
169 to 172

Raw scores	z scores	Z scores
14	1.68	67
13	1.25	63
12	.83	58
11	.41	54
10	− .02	50
9	− .44	46
8	− .86	41
7	−1.29	37
6	−1.71	33
5	−2.14	29
4	−2.56	24
3	−2.99	20

(m) skewness—negative
The mean is less than the median

104 Answers for Computational Mastery Test

3. (a)

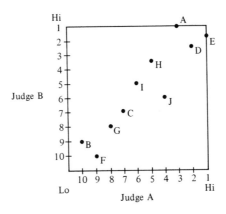

(b) Yes. Positive
(c) +.92

4. (a) $\overline{X} = 8$, $\overline{Y} = 10$

$$r = \frac{125}{\sqrt{(118)(156)}} = \frac{125}{\sqrt{18408}} = \frac{125}{135.68} = +.92$$

(b) $$r = \frac{765 - (64)(80)/8}{\sqrt{(630 - 64^2/8)(956 - 80^2/8)}} = \frac{125}{\sqrt{(118)(156)}} = \frac{125}{135.68} = +.92$$

Appendix 2

CHECK YOUR UNDERSTANDING

Use this to check your understanding and as a learning experience.

1.
Raw scores	Ranks
10	10
15	15
?	20
25	25

    What raw score has been omitted?

    (a) 16, 17, 18, or 19
    (b) 20
    (c) 21, 22, 23, or 24
    (d) impossible to determine from the information given

2.
Raw scores	z scores
60	+2.00
55	+1.00
50	0.00
?	−1.00
40	−2.00

    What raw score has been omitted?

    (a) 41, 42, 43, or 44
    (b) 45
    (c) 46, 47, 48, or 49
    (d) impossible to determine from the information given

3. In a self-contained elementary school classroom, a pupil has a rank of 15 in social studies and a rank of 15 in science.

   (a) This pupil's relative rank is the same in each class.
   (b) This pupil's relative rank is higher in social studies.
   (c) This pupil's relative rank is higher in science.
   (d) It is impossible to compare the relative rank of this pupil for these classes with this information.

4. A pupil has a rank of 23 in physics class. This pupil's rank in typing class is 23.

   (a) This pupil's relative rank is the same in each class.
   (b) This pupil's relative rank is higher in the physics class.
   (c) This pupil's relative rank is higher in the typing class.
   (d) It is impossible to compare the relative rank of this pupil for these classes with this information.

5. A rank of 8 can best be interpreted by comparing the 8 to

   (a) the percentile rank of 100.
   (b) the number in the rank distribution.
   (c) the rank in a second unrelated distribution.
   (d) the median and/or the mean of the distribution.

6. A score on a continuum represents an interval. The score of 77 best represents which of the following intervals?

   (a) 75.5 to 80.5
   (b) 75   to 80
   (c) 74.5 to 79.5
   (d) 74   to 79

7. Q_2 is to the median as Q_3 is to

   (a) P_{25}
   (b) P_{50}
   (c) P_{75}
   (d) P_{100}

8. A test is administered to 150 students. Assuming normal distribution, about how many of these students would have z-scores above +1.00?

   (a) 16
   (b) 25
   (c) 50
   (d) impossible to determine with these data

9. What is the z-score for a raw score that is one and one-fifth standard deviations below the mean?

   (a) −1.50
   (b) −1.20
   (c) +1.50
   (d) impossible to determine with the provided information

10. A distribution of raw scores has a mean of 76 and a standard deviation of 8. About what percentage of scores will fall below a raw score of 84?

    (a) 68%
    (b) 76%
    (c) 84%
    (d) 92%

11. Given: $\overline{X}$ = 42   Median = 42 N = 66   st. dev. = 6

    About how many students scored between 36 and 48 on this test?
    (a) 44
    (b) 33
    (c) 22
    (d) 11

12. What Z-score would be associated with a raw score that is one standard deviation above the raw score mean?

    (a) 50
    (b) 60
    (c) 70
    (d) 80

13. A raw score on a test is 40. The reported Z-score for this case is also 40. The mean for the raw score distribution is

    (a) 50.
    (b) greater than 40.
    (c) 40.
    (d) less than 40.

14. Given: Mean of test A = 72.6
             Mean of test B = 39.6

    John scored 75 on test A and 42 on test B. Considering John's positions in the two distributions, in which testing did he do better?

    (a) better on test A
    (b) no difference
    (c) better on test B
    (d) impossible to determine with this information

15. Given: $N = 99$ $\overline{X} = 47.9$ Median = 52.3

What is the shape of the frequency distribution for these data?

(a) normal (bell-shaped)
(b) negatively skewed
(c) positively skewed
(d) impossible to determine with this information

16. What statistic would be most appropriate to determine the relationship of scores on a first-grade reading readiness test and scores on a test of reading achievement given at the end of grade two?

(a) mean $\overline{X}$
(b) median
(c) standard deviation
(d) coefficient of correlation

17. How many values are needed for each subject to derive a Pearson Product Moment Correlation Coefficient?

(a) one
(b) two
(c) three
(d) four

18. High scores on one variable are associated with high scores on a second variable. Low scores on the first are associated with low on the second. What coefficient of correlation (r) will result from the Pearson technique?

(a) positive
(b) negative
(c) near 0.00
(d) impossible to determine from the information given

19. What information does the correlation coefficient give?

(a) degree of relationship
(b) direction of relationship
(c) cause of relationship
(d) degree and direction of relationship

20. A coefficient of correlation determined for two variables that plot a line (straight by definition) on a scattergram would be

(a) +1.00.
(b) −1.00.
(c) either −1.00 or +1.00.
(d) impossible to determine with this information.

21. Which of the following correlation coefficients reflects the most relationship?

    (a) +.70
    (b) +.30
    (c) -.70
    (d) -1.00

22.

Student		A	B	C	D	E
Scores	Test X	28	24	22	18	16
	Test Y	14	12	11	9	8

If the scores of five students on two tests are as shown above, what is the derived correlation coefficient between the two tests?

    (a) +1.00
    (b) + .50
    (c) - .50
    (d) -1.00

23. There is a high positive correlation between the number of fires and the number of firemen injured. We can conclude from this information that

    (a) the fires were the cause of the injuries.
    (b) firemen should be more careful.
    (c) fire fighting should be discontinued.
    (d) there is a relationship between the number of fires and the number of firemen injured.

24. What statistic would be appropriate to correlate a teacher's rank of student ability and the student's I.Q. scores?

    (a) Pearson r
    (b) Spearman rho
    (c) standard error of estimate
    (d) a technique for handling curvilinear relationships

25. A correlation between college entrance exam grades and scholastic achievement was reported to be +1.10. On the basis of this what would you tell the university?

    (a) The students who score highest on the examination will make the best students.
    (b) The entrance examination is a good predictor of success.
    (c) The entrance examination is a poor predictor of success.
    (d) The university should hire a new statistician.

26. After several studies, Professor Stat concludes that there is a near zero correlation between body weight and bad tempers. What does this mean?

    (a) No one has a bad temper.
    (b) Everyone has a bad temper.
    (c) Heavy people are jolly.
    (d) A person with a bad temper may be heavy or skinny.

27. If the correlation between body weight and annual income were high and positive, then we could conclude that

    (a) high incomes cause people to eat more food.
    (b) low incomes cause people to eat less food.
    (c) there is a relationship between the two variables.
    (d) high income people spend a greater proportion of their income for food.

28. What proportion of the scores in a normal distribution would be above a z-score equal to +1.75?

    (a) .04
    (b) .46
    (c) .54
    (d) .96

29. What proportion of the scores in a normal distribution would be below a z-score equal to +1.28?

    (a) .10
    (b) .40
    (c) .60
    (d) .90

30. A child is born in Professor Stat's family every year for five years. How will the standard deviation of the five children's ages change as they get older?

    (a) standard deviation will decrease
    (b) standard deviation will stay the same
    (c) standard deviation will increase
    (d) standard deviation will decrease at a decelerating rate

31. Which of the following z-scores is most likely to be a standard score found in a distribution of test scores from Professor Ego's psychology class?

    (a) −4.39
    (b) −3.99
    (c) +1.32
    (d) +3.69

32. You obtain a score of 67 on a test. Which class would you rather be in?

   (a) $\bar{X} = 58$ $\sigma = 10$
   (b) $\bar{X} = 60$ $\sigma = 8$
   (c) $\bar{X} = 63$ $\sigma = 4$
   (d) $\bar{X} = 68$ $\sigma = 3$

33. Given: Mean = 30 $\sigma = 12$

   Denny Decile has a z-score of +.50. What was his original raw score?

   (a) 24
   (b) 27
   (c) 33
   (d) 36

34. Pamela Parameter received a raw score of 30 on an examination that had a mean of 35 and a variance of 4. What is her z-score?

   (a) −2.50
   (b) −1.25
   (c) +1.25
   (d) +2.50

35. If the scores of a distribution are transformed to z-scores, then the variance will equal

   (a) −1
   (b) 0
   (c) +1
   (d) N (for that distribution)

36. What measure of central tendency would be most appropriate to report the most typical color of eyes for a defined population?

   (a) range
   (b) mean
   (c) median
   (d) mode

37. The apparent limits of an interval are 15 and 32. The width of the interval is

   (a) 16.
   (b) 17.
   (c) 18.
   (d) 19.

112   Appendix 2

38. It is possible to determine at a glance from a frequency distribution curve (frequency polygon) the _____ .

   (a) mean
   (b) median
   (c) mode
   (d) all of the above

39. How may a graph be used to mislead the reader (viewer)?

   (a) manipulation of the vertical axis
   (b) manipulation of the horizontal axis
   (c) making the initial entry on the ordinate some value other than zero
   (d) all of the above

40. In what type of distribution might the mean be at the forty-fifth percentile?

   (a) negatively skewed
   (b) positively skewed
   (c) normal distribution
   (d) any one of the above

41. If, in a frequency polygon, there are relatively few frequencies at the right end of the distribution when compared with the frequencies at the left end, the distribution is

   (a) positively skewed.
   (b) negatively skewed.
   (c) platykurtic.
   (d) leptokurtic.

42. Variables in which measurement is always approximate because they permit an unlimited number of intermediate values are classified as

   (a) interval.
   (b) ratio.
   (c) continuous.
   (d) discrete.

ANSWERS

Item no.	Answer	Frame
1.	d	2
2.	b	166
3.	a	8
4.	d	8
5.	b	8
6.	c	38
7.	c	111
8.	b	189
9.	b	196
10.	c	193
11.	a	192
12.	b	169
13.	b	171
14.	d	166
15.	b	97
16.	d	203
17.	b	203
18.	a	208
19.	d	212
20.	c	210
21.	d	208, 209
22.	a	208
23.	d	204
24.	b	216
25.	d	212
26.	d	206
27.	c	204
28.	a	189
29.	d	189
30.	b	137
31.	c	189
32.	c	167
33.	d	166
34.	a	166
35.	c	164
36.	d	60
37.	c	31
38.	c	61
39.	d	36
40.	a	97
41.	a	50
42.	c	24

Appendix 3

Table 1. Areas of the Normal Curve in Terms of x/σ.

(1) z Standard Score ($\frac{x}{\sigma}$)	(2) A Area from Mean to $\frac{x}{\sigma}$	(1) z Standard Score ($\frac{x}{\sigma}$)	(2) A Area from Mean to $\frac{x}{\sigma}$	(1) z Standard Score ($\frac{x}{\sigma}$)	(2) A Area from Mean to $\frac{x}{\sigma}$
0.00	.0000	0.25	.0987	0.50	.1915
0.01	.0040	0.26	.1026	0.51	.1950
0.02	.0080	0.27	.1064	0.52	.1985
0.03	.0120	0.28	.1103	0.53	.2019
0.04	.0160	0.29	.1141	0.54	.2054
0.05	.0199	0.30	.1179	0.55	.2088
0.06	.0239	0.31	.1217	0.56	.2123
0.07	.0279	0.32	.1255	0.57	.2157
0.08	.0319	0.33	.1293	0.58	.2190
0.09	.0359	0.34	.1331	0.59	.2224
0.10	.0398	0.35	.1368	0.60	.2257
0.11	.0438	0.36	.1406	0.61	.2291
0.12	.0478	0.37	.1443	0.62	.2324
0.13	.0517	0.38	.1480	0.63	.2357
0.14	.0557	0.39	.1517	0.64	.2389
0.15	.0596	0.40	.1554	0.65	.2422
0.16	.0636	0.41	.1591	0.66	.2454
0.17	.0675	0.42	.1628	0.67	.2486
0.18	.0714	0.43	.1664	0.68	.2517
0.19	.0753	0.44	.1700	0.69	.2549
0.20	.0793	0.45	.1736	0.70	.2580
0.21	.0832	0.46	.1772	0.71	.2611
0.22	.0871	0.47	.1808	0.72	.2642
0.23	.0910	0.48	.1844	0.73	.2673
0.24	.0948	0.49	.1879	0.74	.2703

Table 1. Continued

(1) z Standard Score ($\frac{x}{\sigma}$)	(2) A Area from Mean to $\frac{x}{\sigma}$	(1) z Standard Score ($\frac{x}{\sigma}$)	(2) A Area from Mean to $\frac{x}{\sigma}$	(1) z Standard Score ($\frac{x}{\sigma}$)	(2) A Area from Mean to $\frac{x}{\sigma}$
0.75	.2734	1.20	.3849	1.65	.4505
0.76	.2764	1.21	.3869	1.66	.4515
0.77	.2794	1.22	.3888	1.67	.4525
0.78	.2823	1.23	.3907	1.68	.4535
0.79	.2852	1.24	.3925	1.69	.4545
0.80	.2881	1.25	.3944	1.70	.4554
0.81	.2910	1.26	.3962	1.71	.4564
0.82	.2939	1.27	.3980	1.72	.4573
0.83	.2967	1.28	.3997	1.73	.4582
0.84	.2995	1.29	.4015	1.74	.4591
0.85	.3023	1.30	.4032	1.75	.4599
0.86	.3051	1.31	.4049	1.76	.4608
0.87	.3078	1.32	.4066	1.77	.4616
0.88	.3106	1.33	.4082	1.78	.4625
0.89	.3133	1.34	.4099	1.79	.4633
0.90	.3159	1.35	.4115	1.80	.4641
0.91	.3186	1.36	.4131	1.81	.4649
0.92	.3212	1.37	.4147	1.82	.4656
0.93	.3238	1.38	.4162	1.83	.4664
0.94	.3264	1.39	.4177	1.84	.4671
0.95	.3289	1.40	.4192	1.85	.4678
0.96	.3315	1.41	.4207	1.86	.4686
0.97	.3340	1.42	.4222	1.87	.4693
0.98	.3365	1.43	.4236	1.88	.4699
0.99	.3389	1.44	.4251	1.89	.4706
1.00	.3413	1.45	.4265	1.90	.4713
1.01	.3438	1.46	.4279	1.91	.4719
1.02	.3461	1.47	.4292	1.92	.4726
1.03	.3485	1.48	.4306	1.93	.4732
1.04	.3508	1.49	.4319	1.94	.4738
1.05	.3531	1.50	.4332	1.95	.4744
1.06	.3554	1.51	.4345	1.96	.4750
1.07	.3577	1.52	.4357	1.97	.4756
1.08	.3599	1.53	.4370	1.98	.4761
1.09	.3621	1.54	.4382	1.99	.4767
1.10	.3643	1.55	.4394	2.00	.4772
1.11	.3665	1.56	.4406	2.10	.4821
1.12	.3686	1.57	.4418	2.20	.4861
1.13	.3708	1.58	.4429	2.30	.4893
1.14	.3729	1.59	.4441	2.40	.4918
1.15	.3749	1.60	.4452	2.50	.4938
1.16	.3770	1.61	.4463	2.60	.4953
1.17	.3790	1.62	.4474	2.70	.4965
1.18	.3810	1.63	.4484	2.80	.4974
1.19	.3830	1.64	.4495	2.90	.4981
				3.00	.4987

Table 2. Squares and Square Roots of Numbers from 1 to 120.

N	N^2	$\sqrt{N}$	N	N^2	$\sqrt{N}$	N	N^2	$\sqrt{N}$
1	1	1.0000	41	1681	6.4031	81	6561	9.0000
2	4	1.4142	42	1764	6.4807	82	6724	9.0554
3	9	1.7321	43	1849	6.5574	83	6889	9.1104
4	16	2.0000	44	1936	6.6332	84	7056	9.1652
5	25	2.2361	45	2025	6.7082	85	7225	9.2195
6	36	2.4495	46	2116	6.7823	86	7396	9.2736
7	49	2.6458	47	2209	6.8557	87	7569	9.3274
8	64	2.8284	48	2304	6.9282	88	7744	9.3808
9	81	3.0000	49	2401	7.0000	89	7921	9.4340
10	100	3.1623	50	2500	7.0711	90	8100	9.4868
11	121	3.3166	51	2601	7.1414	91	8281	9.5394
12	144	3.4641	52	2704	7.2111	92	8464	9.5917
13	169	3.6056	53	2809	7.2801	93	8649	9.6437
14	196	3.7417	54	2916	7.3485	94	8836	9.6954
15	225	3.8730	55	3025	7.4162	95	9025	9.7468
16	256	4.0000	56	3136	7.4833	96	9216	9.7980
17	289	4.1231	57	3249	7.5498	97	9409	9.8489
18	324	4.2426	58	3364	7.6158	98	9604	9.8995
19	361	4.3589	59	3481	7.6811	99	9801	9.9499
20	400	4.4721	60	3600	7.7460	100	10000	10.0000
21	441	4.5826	61	3721	7.8102	101	10201	10.0499
22	484	4.6904	62	3844	7.8740	102	10404	10.0995
23	529	4.7958	63	3969	7.9373	103	10609	10.1489
24	576	4.8990	64	4096	8.0000	104	10816	10.1980
25	625	5.0000	65	4225	8.0623	105	11025	10.2470
26	676	5.0990	66	4356	8.1240	106	11236	10.2956
27	729	5.1962	67	4489	8.1854	107	11449	10.3441
28	784	5.2915	68	4624	8.2462	108	11664	10.3923
29	841	5.3852	69	4761	8.3066	109	11881	10.4403
30	900	5.4772	70	4900	8.3666	110	12100	10.4881
31	961	5.5678	71	5041	8.4261	111	12321	10.5357
32	1024	5.6569	72	5184	8.4853	112	12544	10.5830
33	1089	5.7446	73	5329	8.5440	113	12769	10.6301
34	1156	5.8310	74	5476	8.6023	114	12996	10.6771
35	1225	5.9161	75	5625	8.6603	115	13225	10.7238
36	1296	6.0000	76	5776	8.7178	116	13456	10.7703
37	1369	6.0828	77	5929	8.7750	117	13689	10.8167
38	1444	6.1644	78	6084	8.8318	118	13924	10.8628
39	1521	6.2450	79	6241	8.8882	119	14161	10.9087
40	1600	6.3246	80	6400	8.9443	120	14400	10.9545

GLOSSARY

$\| \quad \|$	absolute value
D	difference
$D_{\text{subscript}}$	decile (D_7—seventh decile)
f	frequency (f_b—frequency below)
	(f_w—frequency within)
i	interval
N	number of values
P	percentile
Q	quartile
r	correlation coefficient
ρ	rho (correlation coefficient)
σ	sigma; standard deviation
σ^2	sigma squared; variance
Σ	sigma (capital); sum
$\overline{X}$	mean
z	standard score
Z	standard score

ABSCISSA. The horizontal dimension line on a graph.

ABSOLUTE VALUE. A value that disregards the algebraic sign. |-14| is treated as 14. (Frame 128)

AVERAGE DEVIATION OR MEAN DEVIATION. The arithmetic mean of a set of deviation scores viewed as absolute deviations. (Frame 128)

CORRELATION. A measure of the relationship between two variables. (Frame 203)

CUMULATIVE FREQUENCY POLYGON. The cumulative frequency table values in graphic form. (Frame 43)

CUMULATIVE PERCENTAGE FREQUENCY POLYGON. The cumulative percentage frequency table values in graphic form. (Frame 44)

DECILE. One of nine points that divide a distribution of scores into ten parts that contain the same proportion (.10) of the total number of scores. (Frame 108)

DEVIATION. $X - \overline{X}$. A value that expresses the difference of a score from the distribution's mean.

DEVIATION SCORE. A value derived by subtracting the mean from a raw score. (Frame 122)

FREQUENCY. Number of times a value appears in a distribution.

FREQUENCY DISTRIBUTION. An organization of data into a table or graph using the number of occurrences for each value. (Frame 11)

FREQUENCY POLYGON. A graph that consists of a series of lines connecting points associated with abscissa values. (Frame 39)

GRAPH. A visual representation of a frequency table. Used to give a quick overall picture of a set of data.

118 Glossary

HISTOGRAM.   A graph that represents the frequencies by areas in the form of bars. (Frame 34)

KURTOSIS.   The flatness or peakedness of a distribution. A distribution is classified as platykurtic (flat), leptokurtic (peaked), or mesokurtic (neither flat nor peaked). (Frame 52)

LIMITS.   The endpoints of intervals. (Used as *upper limit* or *lower limit*.) (Frame 24)

MEAN. $\bar{X}$, or sometimes M. The simple arithmetic average. (Frame 83)

MEASUREMENT.   The quantification of a variable by assignment of a meaningful number value.

MEDIAN.   A point in a distribution that divides the number of values (N) into two equal parts (50% and 50%). (Frame 64)

MODE.   The most frequently occurring score in a distribution. (Frame 58)

NORMAL CURVE.   A bell-shaped frequency polygon that is used as a model to study proportions and number of cases for points on the abscissa of a frequency polygon. The limiting form of the binomial as N approaches infinity. (Frame 177)

OGIVE.   The characteristic shape associated with frequency polygons where values do not decrease with larger values on the abscissa. (Frame 43)

ORDER.   Procedure of listing scores from lowest to highest or highest to lowest. (Frame 1)

ORDINATE.   Vertical dimension line on a graph.

PERCENTILE.   One of ninety-nine points used to divide a distribution of scores into 100 equal proportions (.01). (Frame 110)

PERCENTILE RANK.   The percentage of scores in a distribution that fall below a given point or score. (Frame 118)

QUARTILE.   One of three points in a distribution of scores that divide the distribution into four equal proportions (.25). (Frame 107)

r.   Pearson correlation coefficient. The sample value of an estimate of the relationship between two variables. (Frame 221)

RANGE.   Difference between the lowest and highest distribution values. Sometimes defined as $[(X_h - X_l) + 1]$. (Frame 15)

RANK.   A position number assigned to data. (Frame 7)

rho.   Spearman correlation. An estimate of the relationship between two variables that utilizes ranked values for the variables. (Frame 216)

SCATTERGRAM.   A visual representation of the relationship between sets of values on two variables. Constructed by plotting a point for each pair of values obtained from a common subject. (Frame 207)

SKEWNESS.   A distortion or unbalancing within a set of data. Caused by deviate or atypical scores. (Frame 50)

STANDARD DEVIATION. A measure of the variation of a set of scores obtained by taking the square root of the variance. (Frame 139)

VARIABLE. A characteristic or property. Used to describe subjects being studied. Values on the characteristic being studied indicate differences on subjects one from another. (Examples of variables: height, weight, creativity)

VARIANCE. The mean (average) square of the squared deviation scores. (Frame 136)